A HISTORY OF THE
M·I·N·D

A HISTORY OF THE

M·I·N·D

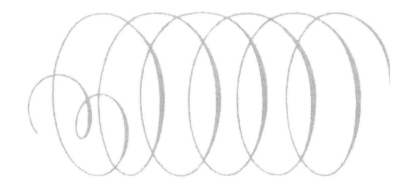

Evolution and the Birth of Consciousness

NICHOLAS HUMPHREY

COPERNICUS

AN IMPRINT OF SPRINGER-VERLAG

First Copernicus softcover edition, 1999.

Published in the United States by Copernicus, an imprint of
Springer-Verlag New York, Inc.

Copernicus
Springer-Verlag New York, Inc.
175 Fifth Avenue
New York, NY 10010

Library of Congress Cataloging-in-Publication Data
Humphrey, Nicholas.
    A history of the mind : evolution and the birth of consciousness /
Nicholas Humphrey.
        p.   cm.
    Originally published: New York : Simon & Schuster, c1992.
    Includes bibliographical references and index.
    ISBN 0-387-98719-3 (softcover : alk. paper)
    1. Consciousness—History.   2. Mind and body—History.   3. Senses
and sensation—History.   4. Genetic psychology.   I. Title.
[BF311.H777   1999]
128′.2—dc21   99-13254

1992 Originally published in the United Kingdom by Chatto & Windus.
1992 First United States edition published by Simon & Schuster.

Cover image: Frances Shroeder/Superstock.
Manufactured in the United States of America.
Printed on acid-free paper.

9 8 7 6 5 4 3 2 1

ISBN 0-387-98719-3   SPIN 10707963

FOR AYLA

CONTENTS

ACKNOWLEDGMENTS

I *have reason* to thank many people for their help: in particular, Peter Bieri, Robert van Gulick, Nicolas Grahek, Ray Jackendoff, Marcel Kinsbourne, Ayla Kohn, Anthony Marcel, Jay Rosenberg, David Rosenthal, and Eckart Scheerer.

But there is one to whom I owe so much more than to any other that his name must stand alone. Daniel Dennett has been the kind of colleague whom everyone hopes for and almost no one gets: a patron, teacher, critic, co-adventurer, and friend. He encouraged me to start this book, gave me a base to do it from, scotched my doubts, raised others, and provided detailed criticism all along the way. Given Dennett's own well-known position on a range of the issues that I cover here and on which he and I remain partly at odds, he may sometimes have thought he had introduced a cuckoo to his nest. So, all the more thanks to him.

In the course of writing, I held a Visiting Fellowship at Dennett's Center for Cognitive Studies in the Department of Philosophy, Tufts University, and subsequently was a member of the "Mind and Brain Group" at the Center for InterDisciplinary Research (ZiF), University of Bielefeld. At a time when Britain has been making academic gypsies of us all, I am especially grateful to these foreign universities for accommodating me. For additional financial and material help I am indebted to the Kapor Foundation (which supported the Fellowship at Tufts), Alec Horsley, and to my publishers, and editor, Jenny Uglow.

READ ME

The *indefinite* article is not without its uses. While it would have been wrong to call this book "*The* History of the Mind," I can call it "A History" without compunction. It is a partial history of a part of what constitutes the human mind: an evolutionary history of how sensory consciousness has come into the world and what it is doing there. But evolutionary history is the biggest part of history and sensory consciousness the best part of the mind.

There have been not a few—perhaps too many—books on mind, consciousness, and evolution published in the last few years (two of them by me). And, as shelves sag and appetites fade, I should explain what is different about this one.

It is different in being more old-fashioned than most. It has very little to say about computers, or artificial intelligence, or the so-called cognitive revolution in psychology. It refers hardly at all to recent developments in the neurosciences. It does not mention quantum theory, or fractals, or morphic fields. It makes no use of sociobiology. In fact, in many respects, this is a book that could have been written a hundred years ago. Only it wasn't. It remains at the cutting edge of theory: but much of the cutting can still be done with a bare spade.

It is different in being more ambitious than most. It sets out not just to define the problem of consciousness but to solve it. After decades of misplaced optimism and subsequent disappointment many scientists and philosophers still see their primary task as being to identify the valley over the next mountain where the rainbow touches earth. But it is time we actually went digging for the crock of gold.

It is different in being about the real thing. Whereas in *Conscious-ness Regained*[1] and *The Inner Eye*,[2] I attempted to explain the nature of "conscious insight" into our feelings, here I return to the nature of feeling as such. Indeed here I totally ignore my earlier position and focus instead on consciousness as raw sensation. When J. M. Keynes was asked by a friend why he was so ready to reject some of his own previous ideas, he replied: "What else do you expect me to do, when I realize I was wrong?" In my own case it is not, I think, so much that I was wrong as that, in my earlier work, I came in at too high a level and left the fundamental problems unresolved.

Other writers about consciousness have tended, as I did previously, to concentrate on second-order mental faculties—"thoughts about feelings" and "thoughts about thoughts." This bias is readily explicable. High-level skills, involving abstract reasoning, language, self-identity, social intelligence, and so on, are signs of human maturity, while raw feelings occur in brutes and babies. The former impress us and surprise us more than do the latter, they seem to require more evolutionary and individual work, they are the perquisites of a grown-up mind—and they are attractive to the theorist. When William Calvin, for example, writes (in another recent book on consciousness): "I really do mean consciousness in the sense of . . . contemplating the past and forecasting the future, planning what to do tomorrow, feeling dismay when seeing a tragedy unfold, and narrating our life story,"[3] or when Roger Penrose writes (in yet another), "it is the ability to divine or intuit truth from falsity in appropriate circumstances—to form inspired judgments—that is the hallmark of consciousness,"[4] I understand their enthusiasm for explaining such remarkable human skills and wish them well. But first things first. Our life story is first of all the story of a *sentient self* or else no story—and this is a book about first things.

I have written the book in the form of a journey of discovery (which duplicates the way my own thinking has gone). The line of reasoning, though not haphazard, is serendipitous—taking advantage, as need arises, of biological evidence here, logical argument there, and pure speculation where nothing else suffices.

Although no author of a theory should hide behind the adage that "the journey not the arrival matters" I do believe that arrivals have very little meaning without journeys. In *The Hitchhiker's Guide to the*

*Galaxy*⁵ it emerges that the solution to the riddle of "life the universe and everything" is "forty-two." Maybe it is. But who cares, if there is no explanation for how or why the answer happens to be 42? By itself, as a bare fact, the answer 42 is merely boring.

Could the solution to the problem of consciousness be boring? Though I say it myself, I suspect that if presented as a bare fact, yes, it could be (perhaps even *should* be). But when the solution is set in an evolutionary context, everything changes.

If I do not greatly delude myself, I have not only completely extricated the notions of Time, and Space . . . but I trust, that I am about to do more—namely, that I shall be able to evolve all the five senses, that is, to deduce them from *one sense* & to state their growth, & the causes of their difference—& in this evolvement to solve the process of Life and Consciousness.

SAMUEL COLERIDGE, *Letter to Thomas Poole,* 1801[6]

· 1 ·

MIND

AND BODY

Everything that is interesting in nature happens at the boundaries: the surface of the earth, the membrane of a cell, the moment of catastrophe, the start and finish of a life. The first and last pages of a book are the most difficult to write.

I am beginning this book on December 25, the anniversary of my father's death. Perhaps I shall finish it by the time my own first child is born.

When my father died, I flew back from America to England and arrived home the next day. He was lying out in his bed in our farmhouse near Cambridge, proverbially asleep. The undertaker came and asked me to show him where the body was. Better, he said, that the family should remain in another room while he and his assistant carried "it" downstairs. The "it," for me, was curiously relieving. My father was no longer there.

For seventy years my father had been a vessel of awareness, a bubble of conscious humanity carried along in the dark foam of insensate matter. For that bounded period he had been a subject to himself, an object to all others. His consciousness was self-contained. What was inside his mind was always outside ours. He had been the center of ideas. He had enjoyed the present tense of raw sensations. He had known what it is like to be a human being. But then, at last, the golden bowl was broken, the bubble burst. From there on, the inside/outside distinction disappeared; or rather there was no inside left to be.

At his funeral we read a passage from *The Pilgrim's Progress* by John Bunyan: "When the day that he must go hence was come, many accompanied him to the river side, into which as he went he said, 'Death where is thy sting?' And as he went down deeper, he said, 'Grave, where is thy victory?' So he passed over, and all the trumpets sounded for him on the other side."[7]

I thought at the same time of the *Cypresse Grove*, by William Drummond: "If two pilgrims, which have wandered some few miles together, have a heart's grief when they are near to part, what must the sorrow be at the parting of two so loving friends and never-loathing lovers, as are the body and the soul?"[8]

There have been serious attempts, even in this century, to observe the "flight of the soul" by scientific measurement. Dr. Duncan Mac-Dougall wrote in the 1907 volume of the *Journal of the American Society for Psychical Research* that he had placed dying patients on a light bed, mounted on a set of carefully balanced scales. He reported sudden weight losses at the time of death of between three eighths and one and a half ounces for six different patients. When he carried out similar experiments with dying dogs, he observed no weight losses at death.[9]

MacDougall's results have not been replicated. When death comes to a person, scarcely an atom need be gained or lost. It is just that the atoms that the person had been made of have been rearranged, and in their new arrangement they no longer constitute a person.

At a church service in Harlem two Sundays back, I heard a black preacher give a sermon on "Taking what is ours." The question, he said, is "Is you is, or is you ain't?" Hamlet put it differently: "To be, or not to be?" It is a question that allows no middle answer. Either it is like something to be oneself, or it is not. A person is, or he ain't. The implications of the *is* are the subject matter of this book.

I have a big fish to fry. But I shall have to spend the first half of the book in catching it; and until I have done so I am reluctant to make any great claims about its size or weight. Its shape, however, I can tell you right away. It has the shape of the Mind–Body Problem.

The mind–body problem is the problem of explaining how states of consciousness arise in human brains. More specifically (and I shall have to be more specific in due time) it is the problem of explaining how subjective feelings arise in human brains.

The vocabulary I have to work with may not serve me well. "Subjective feeling," already, is too vague a term. It is however the term commonly used, even in relatively technical discussions by philosophers, to capture the sense of what it is like to experience consciousness from the inside. Examples of subjective feelings are the sensed redness of a rose, the feeling of a shiver down one's spine, the taste of Roquefort cheese.

Each of us experience such feelings in the "privacy" of our own consciousness, or so it seems. Their "quality" is transparent to us, although it is not something we could easily communicate to someone else; and because quality is so important, indeed intrinsic to the feeling, philosophers sometimes refer to subjective feelings simply as "qualia." No one doubts that subjective feelings have quantitative aspects too: I might be able to tell you, for example, that one sensation of red was *twice* as intense as another. But what I could not tell you (if you did not already know) would be wherein the quality of redness lies.

Now here is the problem, as it emerges from three obvious facts of human life:

Fact 1 is the fact that when, for example, I bite my tongue I experience the subjective feeling of pain (and to remind myself of what that means, I am doing it now). This experience exists for me alone; and were I to try to tell you what it is like, I could do so only in the vaguest and most metaphorical of ways. My felt pain has an associated time (right now), an associated place (my tongue), an intensity (mild), and an affective tone (unpleasant), but in most other respects it seems beyond the scope of physical description. Indeed my pain, I would say, is not a part of the objective world, the world of physical material. In short it can hardly count as a physical event.

Fact 2 is the fact that at the same time as I bite my tongue there are related processes occurring in my brain. These processes comprise the activity of nerve cells. In principle (though not of course in practice) they could be observed by an independent scientist with access to the interior of my head; and were he to try to tell another scientist what my brain-based pain consists in, he would find the objective language of physics and chemistry entirely sufficient for his purpose. For him my brain-based pain would seem to belong nowhere else than in the world of physical material. In short it is nothing other than a physical event.

Fact 3 is the fact that, so far as we know, Fact 1 wholly depends on Fact 2. In other words the subjective feeling is brought about by the brain processes (whatever precisely "brought about by" means).

The problem is to explain *how* and *why* and *to what end* this dependence of the nonphysical mind on the physical brain has come about.

It is a problem that has over the centuries filled philosophers with frustration, desperation, almost panic. Three hundred and fifty years ago René Descartes expressed his sense of helplessness: "So serious are the doubts into which I have been thrown . . . that I can neither put them out of my mind nor see any way of resolving them. It feels as if I have fallen unexpectedly into a deep whirlpool which tumbles me around so that I can neither stand on the bottom nor swim up to the top."[10]

Descartes's own solution was to deny the obvious implication of Fact 3, and to plump for the hypothesis of dualism. Dualism asserts that the universe contains two very different kinds of stuff, mental stuff (of which subjective feelings are made) and physical stuff (of which brains are made), and that these exist semi-independently of one another. Thus in principle there could be minds without brains, and brains without minds. If and when these distinct entities meet and interact—as Descartes of course acknowledged that they do—it involves a handshake across a metaphysical divide.

The trouble with dualism is that it explains both too much and too little, and few philosophers have felt comfortable with it. More recently they have embraced various forms of monism. Monism asserts that there is in reality only one sort of stuff, of which both minds and brains are ultimately made. And in its most extreme form, physicalism, it claims that particular subjective feelings are actually identical to particular physical brain processes (in the same way that a bolt of lightning is identical to an electrical discharge in the air).

Few feel comfortable with this either. It would imply, for a start, that only carbon-based living organisms like ourselves (with carbon-based brains) could have conscious feelings anything like ours. And philosophers have been loath to deny consciousness in advance to other kinds of life forms with differently constituted brains. It would seem chauvinist, to say the least, to suppose that if humanoid creatures had evolved on a faraway planet, using different elements as building blocks, these individuals could have none of the subjective feelings we do—no matter how intelligently and sensitively they behaved. It might be true that they could not, but the truth is certainly not self-evident.

In any case, even if subjective feelings are as a matter of fact identical to physical states, this matter of fact would still cry out for explanation. If we were simply to acknowledge the identity we would have done nothing to dispel the sense of mystery about how it comes to be so. Analogies with lightning bolts would not help either. For in the case of lightning there really is no mystery: any competent physicist could predict that an electrical discharge in the atmosphere would under appropriate conditions produce the flash and bang. By contrast, no one could even begin to predict that the electrical activity of a brain would produce the subjective feeling of tasting cheese.

Samuel Johnson wrote in *Rasselas* in 1759: "Matter can differ from matter only in form, bulk, density, motion and direction of motion: to which of these, however varied or combined, can consciousness be annexed? To be round or square, to be solid or fluid, to be great or little, to be moved slowly or swiftly one way or another, are modes of material existence, all equally alien from the nature of cogitation."[11] And for many modern commentators the same anxieties persist. Colin McGinn, the British philosopher, has written recently: "Somehow, we feel, the water of the physical brain is turned into the wine of consciousness, but we draw a total blank on the nature of this conversion. Neural transmissions just seem like the wrong kind of materials with which to bring consciousness into the world. . . . The mind–body problem is the problem of understanding how the miracle is wrought."[12]

McGinn's unhappy conclusion is that the problem is probably unsolvable: either there actually is not a solution, or, if there is one, human intelligence must always be too limited to grasp it.

Some kinds of problem are unsolvable in principle. There is no solution, for example, to the problem of how to get a quart into a pint pot, or to fit a left hand to a right-hand glove, or (as it happens) to turn water into wine. If the mind–body problem were that kind of a problem, there would be little point pursuing it.

But before we draw any such analogy, we should note an interesting difference between the problems of getting a quart into a pint pot and of getting consciousness into the brain: which is that, while the former has never been known to occur, the latter occurs all the time.

If the conversion of the water of the physical brain into the wine of consciousness is a miracle, it is one of those everyday miracles to which the word "miracle" by definition should not apply.

That being so, we should be careful how we set up the mind–body problem, lest without realizing it we render it not just a difficult problem but one that appears logically intractable.

Gottfried Leibniz in his *Monadology* in 1714 imagined someone walking around a brain, as a factory inspector might walk around a flour mill: "It must be confessed, moreover, that *perception* and that which depends on it *are inexplicable by mechanical causes, that is, by figures and motions.* And, supposing that there were a machine so constructed as to think, feel and have perception, we could conceive of it as enlarged and yet preserving the same proportions, so that we might enter into it as into a mill. And this granted, we should only find on visiting it, pieces which push one against another, but never anything by which to explain perception."[13]

The metaphor is a compelling one, but if you think about it, it has an obvious flaw. Leibniz took the mill to exemplify the bottom line of physical reality. But he might have used the same example to quite different effect. For note that a mill is not simply a physical object. Most importantly it is a *mill*, a machine for grinding grain to yield flour for bread; it is a place of employment; it is a source of wealth. In fact, for the Miller of Dee in the song: "I live by my mill, she is to me both parent, child and wife." Someone who visited the mill and found only pieces which push against each other would not be able to explain any of those properties either. But then that would be because the visitor was falling into the commonsense trap of assuming that the first way something strikes him must be all there is to it: he would be using the wrong level of description.

I once gave a class lecture in which I brought with me a box with two things in it. I drummed on them with a ruler, rat-a-tat-tat. I asked the students to guess what was in the box. "Hollow objects." I let them have a glimpse. "Bones." "Human skulls." One was smaller than the other. "Man and woman's skulls." Taking the skulls from the box, I explained that these were the skulls of American Indians, stolen from a grave. "Put them back." I explained that they were probably man and wife, a young couple who had died together and been buried together; I gave them names and I placed the faces cheek to cheek, Hiawatha and Minnehaha. "That's horrible . . ."

The lesson was that a couple of hollow objects made chiefly of lime can also at another level of description be the relics of two lovers; moreover that what someone does with them can be either a casual entertainment or a gross insult. Different levels of description need have nothing much in common.

Now, what is true for a mill or a skull is certain to be even more true for a highly evolved functional mechanism like a brain. In one sense brains are unquestionably physical objects, which can be described reductively in terms of their material parts. But that is surely not the only way of representing them, nor is it necessarily the most revealing way. What may be required, in order to provide a better clue to how mental activity comes into being, is a way of representing what the brain does over time as distinct from what it is from moment to moment.

One possibility, for example, would be to think of brains as computing machines or logic engines, so that the properties they have for us are not so much physical as mathematical. Thus a brain could be characterized as a device that takes in "information" and "processes" it to yield further information (it certainly does do that, if that is how we choose to describe it); and it could be said that what matters is the mathematical relation between input and output. In that case particular subjective feelings would be identical not to particular physical brain processes but rather to the particular logical operations being performed.

The theory that mental states in general are nothing other than mathematically defined computational states has come to be known as *functionalism*. It has been taken up enthusiastically by several contemporary philosophers. William Lycan, for example, wrote in a recent book that this is "the only positive doctrine in all of philosophy that I am prepared (if not licensed) to kill for."[14] But while many others would agree that there could be an equivalence between computational states and certain sorts of mental processes, they draw the line at conscious mental processes, and draw it more firmly still at conscious awareness of subjective feelings.

It is certainly a strange idea: that states of consciousness correspond to logical rather than material states of the brain. It seems especially strange when we realize that, if it is right, these same logical states could exist in an inanimate machine and that the machine (whatever it was made of) would thereby have conscious feelings.

The idea is too strange for some. To quote McGinn again: "You can't get the 'qualitative content' of conscious experience—seeing red, feeling a pain, etc.—out of computations in the nervous system."[15] Or to quote Ray Jackendoff, author of *Consciousness and the Computational Mind*: "I find it every bit as incoherent to speak of conscious experience as a flow of information as to speak of it as a collection of neural firings."[16]

Maybe, however, it is just that we do not yet know enough about the nature of the thing that the nervous system must be computing, and when we do it will not seem such a miracle.

Well, we shall see . . . But not before we have a better fix on what the "mind" side of the mind–body problem is. And this will require a major rethink—or repair—of widely held assumptions about what minds are *for*. Although my aim is indeed to explain "consciousness" in sentient human beings, there is much that has first to be said about being human, and before this much that has to be said about being sentient.

· 2 ·

"PUZZLING WORK":
AN ASIDE ABOUT
LANGUAGE

Though I have hardly begun, I want to stop and make some prophylactic comments about the use of words. In what I have already written, and more so in what is to come, several of the key terms are put in quotation marks or emphasized, a sure sign that the words in question are not quite right. Sometimes, as J. Alfred Prufrock laments in T. S. Eliot's poem, it seems that:

> It is impossible to say just what I mean!
> But as if a magic lantern threw the nerves in patterns on a screen.[17]

Yet if it is true that our linguistic resources for talking about the mind are so poorly developed, this might be taken to imply that there is something seriously wrong with the whole enterprise. After all, human beings have been talking around and about these questions for a long, long time. If it is still so hard to find the right words to describe such seemingly essential notions as mind and consciousness, perhaps it means that these notions are not so essential after all.

There is a strong tradition in twentieth-century philosophy to the effect that if and when we cannot say exactly what we mean, we probably do not mean anything worth saying. "Anything that can be said," Ludwig Wittgenstein wrote, "can be said clearly." But the situation is not really so clear-cut. Wittgenstein's Cambridge colleague C. D. Broad claimed that "clarity is not enough." He meant that speaking clearly is no guarantee of speaking sense—that clarity, even if it is necessary, is not sufficient. But maybe total clarity is not necessary either. As we all know, a great many things that human beings actually do say to each other are not said clearly. And yet, it seems, we succeed in conveying most of what we want to convey most of the time.

We should not take a Panglossian view of human language. Dr. Pangloss's maxim was that "All is for the best, in the best of all possible worlds." He would no doubt have considered that everything about our language is already as good as it could possibly be. But he would surely have been wrong. For just as an individual child has to acquire a vocabulary in the course of growing up, so does a human culture; and it may well be that in some areas of discourse our linguistic culture is still at the infant stage.

A revealing example of linguistic immaturity occurs with Plato, who it seems had great difficulty in speaking about numbers. In *The Republic* Socrates is discussing how the Guardians of the state should organize a breeding program for the citizens: "Though the Rulers you have trained for your city are wise, reason and perception will not always enable them to hit on the right and wrong times for breeding; some times they will miss them and then children will be begotten amiss." Fortunately, says Socrates, it can all be worked out by arithmetic: "For the human creature the number [of gestation] is the first in which root and square multiplications (comprising three dimensions and four limits) of basic numbers, which make like and unlike, and which increase and decrease, produce a final result in completely commensurate terms."[18]

If this is all Greek to you, you are in good company, for even early classical commentators could not make out what was meant. It is now generally agreed that the number in question—"Plato's number"—was 216; and 216 days is 7 months, which was regarded by the Greeks as the minimum period of gestation (normal gestation being calculated as $216 + 3 \times 4 \times 5 = 276$).

Now, 216 is 6 cubed, and it is also equal to 3 cubed + 4 cubed + 5 cubed. It was this property that Plato was apparently trying to specify. But although he must have known about "cubing"—have understood intuitively its mathematical significance—he did not have a word for it. And the best he could do, so scholars suggest, was to employ the clumsy expression "root and square multiplications (comprising three dimensions and four limits)."

It may seem to us now almost bizarre that Plato of all people should have been lost for a way of expressing such a simple concept as "taking the third power." Every modern schoolchild can do better. But, however that may be, no one presumably would want to claim that Plato's linguistic embarrassment could ever have implied that "cubing" was—or is—an idea about which it would have been better not to have said anything at all.

The lesson I would draw is that perhaps we ourselves are now in the same position with respect to the language we have at our disposal for talking about mind and consciousness. At this stage in our cultural development there are still things we can appreciate intuitively which as yet we have no good way of putting into words.

The problem becomes especially obvious when one national language has resources that another lacks. There is a famous essay by the philosopher Thomas Nagel that is entitled "What Is It Like to Be a Bat?"[19] In French this has been translated (with an apologetic note from the translator) as "Quel effet cela fait d'être une chauve-souris?"[20]—literally "What effect (or impression) does it make to be a bat?" Since the point of Nagel's essay is to argue precisely that the subjective experience of a bat cannot be described in terms of its observable *effects*, there would seem to be a real danger that French readers will not entirely get his message. Yet who doubts that French speakers have the concept—if only it can be addressed—of what in English we express as "what it's like to be . . ."?

This is one of the problems with language. But there is another that is nearly the opposite. While sometimes we are lost for words, at other times the words come all too easily. The fact that a word or phrase exists in our language and is available for use is no guarantee that it can do a useful job. Certain words are, as it were, impostors, that promise much more than they deliver (in fact there are those who would argue that "what it's like to be . . ." is just such a case!).

One of the best-known examples is the word "phlogiston," coined in the eighteenth century to refer to the hypothetical material with negative mass that was supposed to be released from combustible bodies on burning. But we might think too of "élan vital," "animal magnetism," "telepathy," not to mention a host of words with more impressive pedigrees such as "Father Christmas," "the Loch Ness Monster," and "nuclear deterrence."

George Eliot wrote in her journal for 1856: "I have never before longed so much to know the names of things. The desire is part of the tendency that is now growing in me to escape from all vagueness and inaccuracy into the daylight of distinct vivid ideas. The mere fact of naming an object tends to give definiteness to our conception of it."[21] But the mere fact that naming something tends to give definiteness to our conception of it can cut both ways. Once we have a word for something it is easy to suppose that *ipso facto* the thing named is a distinctive entity.

The Great Train Robbery in England in the 1960s provides a comic illustration. The police had made no progress whatever in solving the crime. Eventually the head of Scotland Yard called a news conference where he announced, with evident satisfaction, that he could now reveal that "there was a Brain behind the robbery." His statement evoked a mocking comment from the French newspaper *Le Monde*: "Tout est expliqué. Un Cerveau, c'est quelque chose!" But of course nothing was "expliqué," since the "Cerveau" was not "quelque chose!" at all. Scotland Yard's naming of the Brain was no more than a convenient way of explaining away their inability to catch the robbers.

Taken together, these two problems with language create a kind of double jeopardy for discussions of the mind: there are likely to be certain areas where, as it were, the words play hard to get, and others where they sing a Siren song. One of George Eliot's characters, Mr. Tulliver, put the point nicely in conversation with his wife: "No, no Bessy . . . I meant [what I said] to stand for summat else; but never mind—it's puzzling work, talking is."[22]

To illustrate just how puzzling is the work of talking about mind, consider several recent statements about "consciousness":

"Consciousness is the greatest invention in the history of life; it has allowed life to become aware of itself." [Stephen Jay Gould (biologist)][23]

"Conscious awareness is a conditional property of the reality model in its tripartite form. It may be said to be the subjective aspect of the continuing re-presentation of a temporally stabilized informational display within which multilateral processing of an issue can occur." [John Crook (ethologist)][24]

"In all the contexts in which it tends to be deployed, the term "conscious" and its cognates are for *scientific* purposes both unhelpful and unnecessary." [Kathleen Wilkes (philosopher)][25]

"Reference to consciousness in psychological science is demanded, legitimate, and necessary. It is demanded since consciousness is a central (if not *the* central) aspect of mental life. It is legitimate because there are as reasonable grounds for identifying consciousness as there are for identifying other psychological constructs. It is necessary since it has explanatory value, and since there are grounds for positing that it has causal status." [Anthony Marcel (psychologist)][26]

"I find that I have no clear conception what people are talking about when they talk about 'consciousness' or 'phenomenal awareness.' " [Alan Allport (psychologist)][27]

To which I would add the famous passage from William James, in 1904: " 'Consciousness' . . . is the name of a non-entity, and has no right to a place among first principles. Those who still cling to it are clinging to a mere echo, the faint rumor left behind by the disappearing 'soul,' upon the air of philosophy. . . . It seems to me that the hour is ripe for it to be openly and universally discarded."[28]

James actually went further. "Breath," he wrote, "moving outwards, between the glottis and the nostrils, is, I am persuaded, the essence out of which philosophers have constructed the entity known to them as consciousness." That the man who a few years earlier had popularized the idea of the "stream of consciousness" in his *Principles of Psychology* should have become so hostile to the very term suggests an unusual degree of disillusionment.

Perhaps James would have liked the remark of an American schoolboy, reported in a recent edition of *The Boston Globe*. The boy had been asked to write an essay about vacuums: "Vacuums," he said, "are nothings. We only mention them to let them know we know they're there."[29]

He might have been amused too by the report of a 1960s Loch Ness investigator, Maurice Burton. "From my own experience and those of other observers, there is one statement more true than another: it is that the Loch Ness Monster comes to the surface with surprising infrequency."[30]

After the Loch Ness Monster had supposedly been photographed by an underwater camera, the naturalist Sir Peter Scott suggested in the journal *Nature* that it now merited a scientific name: *Nessiteras rhombopteryx*—Ness dweller with rhomboidal fins.[31] By an unhappy accident, the name was an anagram of "monster hoax by Sir Peter S."

There may well prove to be problems about calling consciousness by name. But they should not prove insuperable. For, if there is one statement that, if not more true than another, is nonetheless true, it is that consciousness comes to the surface with surprising *frequency*.

· 3 ·

WHAT HAPPENED
IN HISTORY:
THE INSIDE STORY

There are several ways to catch a fish (if not a monster). You can drag a net across the river, and pull in everything there is: but this way you get the weeds, the frogs, and old boots too. You can put a worm on a hook, and cast it into a likely-looking pool: but this way you risk choosing the wrong pool or a day when the fish are just not feeding. Or (so an old Scotsman told me) you can tickle it: you walk stealthily along the riverbank until you see your fish hanging in the water just upstream; you lean down from the bank and lower your fingers ever so slowly under the fish's belly; you stroke it; and then (so he said) the fish just lets you lift it out.

I believe the way to catch consciousness will be to tickle it. That is to say we should discover where it is lying, approach it slowly, and then charm it into our hands.

The story line of the book will be a history of mental life. By "history" I mean evolutionary history, and evolutionary history on a grand scale: from the creation of the Earth to the emergence of modern human beings. And the reasons for embracing such a vast time scale are twofold: first, so as to make no preliminary assumptions about when mind and consciousness emerged, and second, so as to make no assumptions about objective physical reality.

Suppose we were to take a relatively shorter time span, say only the last million years. We should then be faced with two existing sets of facts: on the one hand the existing phenomena of subjective experience and on the other the existing phenomena of the material world. The problem then might be precisely the problem we met in the earlier chapter, namely that those two classes of phenomena seem simply not to join.

When we take the longer view, however, we may be able, as it were, to get in on the ground floor before these existing phenomena were phenomena at all. Perhaps we may discover that both classes of phenomena, rather than being "given," are themselves historical creations: the left hand of subjective experience and the right hand of the material world being outgrowths from a common source. In that case the problem will be to trace their separate paths of evolution.

I take it for granted that the human mind does indeed have an evolutionary history, extending through nonhuman prototypes—monkeys, reptiles, worms—all the way back to the first glimmerings of life on Earth. (If, to the contrary, human beings have been the product of all-at-once divine creation, my line of argument would fail; but so would natural philosophy in general.) Before life emerged, let's say four billion years ago, when the planet Earth was formed, there were presumably no minds of any kind at all.

It follows that four billion years ago the world was totally unexperienced and unknown. Nothing within it had ever been seen, heard, touched, smelled, thought about, represented, or described. And hence nothing within it, at that time, existed *as a phenomenon for* anyone. I am, I should say, using the term "phenomenon" here in the old-fashioned way: a "phenomenon" (from the Greek *phainein*, to appear) is an event as it appears to an observer, as distinguished from what it might consist of in itself.

Then, at that stage of our planet's history, the phenomena we now call subjective feelings were not yet in existence: no sensations of red or stabs of pain. Less obviously, although no less true, the phenomena we now call the phenomena of the material world were not yet in existence: no red light or sharp objects, or even objects weighing five pounds or seven feet high—at least nothing that had ever been thought of in that way. I am not making a particularly deep point here: just the point that before anything could exist as a subjective feeling or as a physical event, there had to be someone around to whom that was what it was or meant.

You may object that you cannot imagine a time when nothing existed in any phenomenal form. Were there not volcanoes, and dust storms, and starlight long before there was any life on Earth? Did not the sun rise in the east and set in the west? Did not water flow downhill, and light travel faster than sound? The answer is that if you had been there, that is indeed the way the phenomena would have appeared to you. But you were not there: no one was. And because no one was there, there was not—at this mindless stage of history—anything that *counted as* a volcano, or a dust storm and so on. I am not suggesting that the world had no substance to it whatsoever. We might say, perhaps, that it consisted of "world-stuff." But the properties of this worldstuff had yet to be represented by a mind.

Now, four billion years later, the situation has dramatically changed. Today there are literally billions of animals with minds inhabiting the planet, and the world has become very widely experienced and very widely known. In particular the phenomena both of subjective feelings and of the material world have come into existence as such for us. Today we can go beyond our given interactions and conceive of the existence of comparable phenomena in parts of space where we have never been, and far back in the past and forward in the future. We can imagine the sound of a tree falling in the forest when there is nobody around. We can even imagine, perhaps, the original Big Bang. The fact remains that, whatever the Big Bang was like, there was no phenomenal bang at the time that it occurred.

Having fixed both ends, the big question must be what happened in the period in between.

I shall merely sketch here a possible version of the history, in several acts. (And although, given what I have just said, there must be something paradoxical in using modern concepts to discuss the distant past, this will have to be a contemporary mind's-eye view.) If I seem to move unreasonably quickly past episodes, perhaps whole scenes, that deserve more careful and detailed treatment, I can only ask you to take some of it temporarily on trust.

In the primeval soup, chance brought together the first molecules of life, with the capacity to generate new copies of themselves. Time passed and Darwinian evolution got to work, selecting—and hence helping to design—packets of worldstuff with ever greater potential for maintaining their own integrity and reproducing. First there were just complex living molecules (like DNA), then single cells (like bacteria or amoebae), then multicelled organism (like worms, or fish, or us).

Living animals had their own form and their own substance. Not only was each individual animal a spatially bounded package, but in an important sense the contents of the package belonged together. Although the meaning of "ownership" and "belonging" is intuitively obvious (which tells us how important the idea of "owning" our own bodies remains to our own lives), they are elusive concepts to which I shall return in later chapters. For the moment, however, all I want to imply is that, whether at the level of an amoeba or an elephant, the animal was a self-integrating and self-individuating whole. And unlike other bounded objects—such as a raindrop or a pebble or the moon—its boundaries were self-imposed and actively maintained. On one side of its boundary wall lay "me," on the other "not-me": and it was "my life," "my form," "my substance" that was at risk.

So boundaries—and the physical structures that constituted them, membranes, skins—were crucial. First, they held the animal's substance in, and the rest of the world out. Second, by virtue of being located at the animal's surface they formed a frontier: the frontier at which the outside world impacted the animal, and across which exchanges of matter and energy and information could take place.

Light fell on the animal, objects bumped into it, pressure waves pressed against it, chemicals stuck to it . . . Some of these events were, generally speaking, "a good thing" for the animal, others were neutral, others were bad. Any animal that had the means to sort out the good from the bad—approaching or letting in the good, avoiding or blocking the bad—would clearly have been at a biological advantage. Natural selection was therefore likely to select for "sensitivity."

Being sensitive need have meant, to begin with, nothing more complicated than being locally reactive: in other words, responding selectively at the place where the surface stimulus occurred. Just as today we might say that a person is sensitive to sunlight if he responds to sunlight on his neck with local reddening, so the first types of sensitivity would have involved, for example, local retraction or swelling or engulfing by the skin.

Soon enough, however, more sophisticated types of sensitivity evolved. Sense organs became more discriminatory between different kinds of stimuli, and the range of possible responses increased. Instead of or as well as a stimulus inducing a local reaction, information from one part of the skin got relayed to other parts and caused reactions there. And by the introduction of delays in transmission and the combination of facilitation and inhibition, the way was open for the animal's responses to become better adapted to its needs: for example by swimming away, rather than just recoiling from a noxious stimulus.

In time, different stimuli came to elicit very different action patterns. We might imagine, to take a hypothetical example, that an animal living in a pond swam upward in response to red light, and downward in response to blue light (thus tending to go deeper in the middle of the day). Since information about the particular stimulus was now being preserved and carried through into the particular action pattern, the action pattern had come to represent—at least to replicate symbolically—the stimulus.

With this level of sensitivity and reactivity, however, it could hardly be said that environmental events had acquired much "meaning" for the animal. Still, even by this stage something about the status of the world was changing. Certain events were being responded to *as* good and bad, *as* edible or inedible, *as* of significance to "me." And the reason for emphasizing the *as* here is to emphasize the essential difference between, on the one hand, something's just being good or bad, and, on the other, the animal for whom it is good or bad reacting to it as such. Compare, for example, the effects of low humidity on two bounded objects: a wood louse and a puddle. The heat is "bad" for both of them because it dries them up. But whereas the puddle just sits there and shrinks in size, the wood louse runs away. Both react to low humidity: but while the puddle's response is nonadaptive and carries no implication of being meaningful, the wood louse's response potentially does: it implies "here is a situation not much to my liking."

"Liking" is another of those concepts that I shall want to explore in more detail later on. The question of how much an animal likes being stimulated is, I think, basic to the question of what it is like for the animal to respond to the stimulus (and the pun on "like" is thus not accidental). There are many dimensions and degrees of liking and disliking, corresponding to the many different kinds of sensitivity and responsivity that have evolved. Within this rich space of affective reactions there must have been wide scope for the evolution of ways of experiencing the world that varied in subjective quality.

To begin with, sensitivity and responsivity were intimately linked. And so in some ways they always have been and still are. (Consider, for example, that an itch is something you want to scratch, or that a heavy object is something it is difficult for you to lift.) But as animals became increasingly sophisticated at attuning their behavior to the environmental situation, the sensory side and the response side of the process must have become partially decoupled. Before long a central site evolved, where representations—in the form of action patterns—were held in abeyance before they were put into effect. Thus action patterns had become action plans, and representations had become relatively abstract. The place where they were held in store could be said to be the place where they were held in mind.

"Mind," more than any other term, is embarrassingly difficult to give a simple definition. But, recognizing fully the circularity, I shall let the term "mind" connote for the moment just the representative faculty I have here referred to. In short, animals first had "minds" when they first became capable of storing—and possibly recalling and reworking—action-based representations of the effects of environmental stimulation on their own bodies. The material substrate of the mind was nervous tissue, which in higher organisms became centered in a ganglion or brain; and it is to be remarked that even in animals like human beings the neural tube which forms the brain during embryological development derives from an infolding of the skin.

By the time prototypical minds had evolved, it could be said that some events in the world had taken on the status of *meaningful* phenomena. For the first time in history—the first time in fact since the universe began—certain events, namely those occurring at the surfaces of living organisms, had begun to exist as something for someone. If you will pardon the word play, these events had begun at last to be "matters of fact" because someone "minded" about the fact they "mattered" to his bodily well-being.

So the phenomenology of sensory experiences came first. Before there were any other kinds of phenomena there were "raw sensations"—tastes, smells, tickles, pains, sensations of warmth, of light, of sound, and so on.

It could have happened, I suppose, that this was where mental representation stopped evolving. Indeed it is quite conceivable that somewhere far away in another galaxy where life is evolving on another planet, this is still as far as it has gone; even on Earth it may be as far as some primitive animals have gotten to; it may even correspond to the condition, for a short time, of a newborn human baby. But it is clearly not where our own mental representation rested. For if it were, we would still be living in a world where objective physical phenomena were quite unknown.

From early on there was, however, another track to mental evolution. On the one hand, as we have seen, animals benefited from having an ability to assess their own current state of being: to answer questions about "what is happening to me"—"What is it like to have red light arriving at my skin?" But on the other hand they would certainly have benefited further if they had had an ability to assess the state of the external world: to answer questions about "what is happening out there"—for example, "Where is the red light coming from?" But the questions "What is happening to me?" and "What is happening out there?" were always different kinds of questions, which must always have required very different kinds of answers.

Consider a patch of sunlight falling on the skin of an amoeba-like animal. The light has immediate implications for the animal's own state of bodily health, and for that reason it gets represented as a subjective sensation. But the light also signifies—as we now know— an objective physical fact, namely the existence of the sun. And, although the existence of the sun might not matter much to an amoeba, there are other animals and other areas of the physical world where the ability to take account of what exists "out there beyond my body" could be of paramount survival value. Consider a shadow crossing the skin of the amoeba. Here an ability to represent the objective fact of an approaching predator would—if only it were achievable by an amoeba—clearly be of considerably more consequence to the animal's survival than the ability to represent the body surface stimulus as such.

But how to do it? How to interpret a stimulus as a "sign" of something else? To move from a representation of the sign to a representation of the signified? By the end of the first stage of evolution sense organs existed with connections to a central processor, and most of the requisite information about potential signs was being received as "input." But the subsequent processing of this information, leading to subjective sensory states, had to do with quality rather than quantity, the transient present rather than permanent identity, me-ness rather than otherness. In order that the same information could now be used to represent the outside world, a whole new style of processing had to evolve, with an emphasis less on the subjective present and more on object permanence, less on immediate responsiveness and more on future possibilities, less on what it is like for me and more on how what "it" signifies fits into the larger picture of a stable external world.

To cut a long story short, there developed in consequence two distinct kinds of mental representation, involving very different styles of information processing. While one path led to the qualia of subjective feelings and first-person knowledge of the self, the other led to the intentional objects of cognition and objective knowledge of the external physical world.

When the Earth was formed neither kind of phenomenon existed for anyone at all. Now both exist as such for us. And it is the evolution of these dual modes of representation which goes a long way to explain why now, today, we have this apparent standoff between two classes of phenomena: subjective feelings ranged against the phenomena of the material world, quality against quantity, wine against water. As Picasso said (in a rather different context), "Nature and art being two different things cannot be the same thing";[32] and, by the same token, subjective feelings and physical phenomena, being two different sorts of representation, cannot be the same sort of representation.

· 4 ·

THE DOUBLE
PROVINCE OF
THE SENSES

Having *started* on this evolutionary story, I might be expected to go on at once to greater depths. But since I have been tailoring the story to fit the contemporary facts, I ought first to spend some time on examining rather carefully what these facts are. So let me jump far ahead, to what I take to be the condition of a living human being.

Here I am, sitting at my desk, by a window overlooking a country garden on a summer afternoon, with a cup of hot tea in my hand, the sound of distant rumbling in my ears, and an ant (or something) crawling up my leg. My body surface is being bombarded by environmental stimuli. On one level, just like the primitive amoeba, I am interpreting these stimuli as events that directly affect my bodily state: I like some and I dislike some, and the quality of my liking and disliking varies hugely. At this level, I am at the center of my private world of immediate and direct sensations. On another level, I am interpreting the same surface stimuli as signs, signifying the state of the external world: I see the flowers in bloom, I hear the thunder, I smell the scent of lavender, I think it is an ant, I can tell from the sun's height the time of day. At this second level, I am the spectator of a public world (not *my* world now) of independent physical phenomena.

Admittedly, this way of putting things might be considered no more than that—a "way of putting things," without any special claim to capturing the metaphysical or psychological reality. I would stress therefore that it is a way of putting things that several distinguished writers have settled on before me.

Thomas Reid, leader of the school of Scottish philosophers, wrote in his *Essays on the Intellectual Powers of Man* in 1785, "The external senses have a double province—to make us feel, and to make us perceive. They furnish us with a variety of sensations, some pleasant, others painful, and others indifferent; at the same time they give us a conception of and an invincible belief in the existence of external objects. This conception of external objects is the work of nature; so likewise is the sensation that accompanies it. This conception and belief which nature produces by means of the senses, we call *perception*. The feeling which goes along with perception, we call *sensation*. . . . When I smell a rose, there is in this operation both sensation and perception. The agreeable odour I feel, considered by itself, without relation to any external object is merely a sensation. . . . Perception [by contrast] has always an external object; and the object of my perception, in this case, is that quality in the rose which I discern by the sense of smell."[33]

Sigmund Freud wrote of two principles of mental functioning, the "pleasure" and the "reality" principle. And more recently the psychiatrist Ernest Schachtel distinguished between what he calls the "autocentric" and "allocentric" modes of experiencing the world: "The main differences between the autocentric and allocentric modes of perception are these: In the autocentric mode there is little or no objectification; the emphasis is on how and what the person feels; there is a close relation, amounting to a fusion, between sensory quality and pleasure or unpleasure feelings, and the perceiver reacts primarily to something impinging on him. . . . In the allocentric mode there is objectification; the emphasis is on what the object is like."[34]

But closest of all to the ideas that I have been putting forward are the ramblings of an obscure psychologist called E. D. Starbuck. In a paper titled "The Intimate Senses as Sources of Wisdom," published in the *Journal of Religion* in 1921, Starbuck discussed the distinction between "intimate" and "defining" sensory processes. In the circumstances I think I should quote him at some length:

"In so far as a receptor discriminates qualities in objects and perceives their kinships it may be called a *defining* sense. Since all the senses possess this power to a certain degree it is more fitting to speak of defining sensory processes. . . . Some of the other senses are concerned with the interpretation of objects and of their qualities *immediately* without defining them or setting them into spatial and temporal orders. Their qualities are *directly* regarded as agreeable or indifferent, as desirable or undesirable, or otherwise fitted to the well-being of the organism. In so far as a receptor reports to consciousness directly or immediately qualities of objects together with cues of right response, it may be designated an *intimate* sense. Or again, since all of the senses have in greater or less degree this propensity, it is better to speak of intimate sensory processes. . . . There has been a double line of development and evolution equally important: the one moving fast and far in the direction of description, scientific analysis, practical manipulation, logical construction, and system-building. The other line has achieved equal success in interpreting its objects and their meanings in subtle and skillful ways and in holding the individual in right relationship to his world of experience. . . . Since there is more than one way of interpreting the outer world of experience, the ultimate reason for it may be that there is more than one sort of objective reality."[35]

The claim is that the two categories of experience—sensation and perception, autocentric and allocentric representations, subjective feelings and physical phenomena—are alternative and essentially nonoverlapping ways of interpreting the meaning of an environmental stimulus arriving at the body. So that, when I smell a rose, sensation provides the answer to the question "What is happening to me?" and perception the answer to the question "What is happening out there?"

The distinction, however, is not one that is always apparent in ordinary language. The point was underscored by Reid: "Sensation, taken by itself, implies neither the conception nor belief of any external object. It supposes a sentient being, and a certain manner in which that being is affected; but it supposes no more. Perception implies an immediate conviction and belief of something external—something different both from the mind that perceives, and the act of perception. Things so different in their nature ought to be distinguished. . . . [But] the perception and its corresponding sensation are produced at the same time. In our experience we never find them disjoined. Hence, we are led to consider them as one thing, to give them one name, and to confound their different attributes. It becomes very difficult to separate them in thought, to attend to each by itself, and to attribute nothing to it which belongs to the other."[36]

The term "sweet," for example, can be used both for the subjective sensation I have when a rose's scent reaches my nostrils and for the perceived smell of the rose in its own right; similarly, "red" for the sensation I have when light from the rose's petals reaches my eyes and for the perceived color of those petals, and "sharp" for the sensation I have when its thorns press against my skin and for the perceived shape of those thorns.

If we were to hold to what I called the Panglossian view of language, we might be tempted to conclude that since our familiar vocabulary lumps sensations and perceptions together, they must to all intents and purposes amount to the same thing. But we have only to think of other examples of linguistic lumping to see that such a conclusion would be unwarranted. Consider for example the words that are used to name farm animals and/or the meat that comes from them. In the French language a single word serves for both: *mouton* for both the sheep and sheep meat, *boeuf* for the bullock and bullock meat, *porc* for the pig and pig meat. In English we generally have two words (having retained the Saxon word for the animal while borrowing the Norman French word for the meat)—sheep/mutton, bullock/beef, pig/pork, and so on—but, even so, we use the words lamb or chicken, for example, with both meanings.

Perhaps we should not discount the possibility that one day the English language will have different words to describe sensations and perceptions. At present, however, it is as though we are still at the pre–Norman Conquest stage.

There has been too much philosophical dissension in this area, centered around language, for me to be able to assume that everyone will go along with this distinction without further persuading. But the reality and significance of it will I think be reinforced over the next few chapters. And, for the time being, I want to put the difficulties of language to one side, and to turn to a troubling and important problem, that has also been one of the chief sources of dissent: the question of how sensation and perception, assuming that they are distinct, are causally related.

There are two obvious possibilities. One would be that sensation and perception are independently processed by *parallel* channels of the mind:

rose → chemical odor at nose ⤙ sensation of myself being sweetly stimulated

perception of the rose as having a sweet scent

Or more generally:

object → body surface stimulus ⤙ sensation of what is happening to me

perception of what is happening out there

The other (the theory that in some ways might seem much more plausible) would be that sensation and perception follow on *serially* one from the other:

rose → chemical odor at nose → sensation of myself being sweetly stimulated → perception of the rose as having a sweet scent

Or more generally:

object → body surface stimulus → sensation of what is happening to me → perception of what is happening out there

Reid's own view on this matter was interestingly ambiguous. At one point in his *Essays* he insisted that perception is "immediate" and "not dependent on reasoning," being "a part of the original constitution of the human mind." But then he wrote: "Observing that the agreeable sensation is raised when the rose is near, and ceases when it is removed, I am led by my nature, to conclude some quality to be in the rose, which is the cause of this sensation. This quality in the rose is the object perceived. . . . All the names we have for smells, tastes, sounds, and for various degrees of heat and cold . . . signify both a sensation, and a quality perceived *by means* of that sensation [my emphasis]."[7] By which, presumably, he was implying that perception is secondary to and derived from sensation: in fact that perception is a "conclusion" based upon sensation.

Now, if this latter theory were to be correct, the case I am making would obviously be undermined. It would mean that rather than there being two independently evolved channels of mental representation there would in reality be only a single channel—whose products happen to reach consciousness in, as it were, a relatively unprocessed form, sensation, and a processed form, perception. If this were so, the significance of the distinction between these two categories of experience—and with it the distinction between subjective feelings and physical phenomena—would be lost.

So the question is: is there any conclusive way of deciding which scheme, the parallel or the serial, is right? And the answer lies in examining the possibility of sensation and perception being "decoupled." For it will be apparent that while the parallel scheme would allow for sensation and perception to go their own ways, the serial scheme would not. If perception is causally dependent on sensation, any change in sensation would be bound to have a knock-on effect upon perception; and if there were to be a complete disruption or breakdown in sensation, perception would be wiped out altogether.

I shall be presenting evidence in Chapters 10 to 12 to demonstrate that sensation and perception *can* go their own ways, and indeed that perception can occur in the total absence of sensation: in other words, evidence that there really are two parallel channels in the mind. But this evidence will be much more persuasive if I explore some other issues first.

In the history of psychology, the controversy about whether there is one channel or two raged throughout the nineteenth century. And it had an unfortunate effect. For, as doubts began to arise about whether perception is in fact serially dependent on sensation, many psychologists concerned with sensory processes took to concentrating entirely on perception and stopped being interested in sensation as such at all. And with that, they stopped being interested in "autocentricity," "intimacy," "affect"—and ultimately with the whole area of "subjective feeling."

In 1623 William Drummond could write: "What sweet contentments doth the soul enjoy by the senses. They are the gates and windows of its knowledge, the organs of its delight."[38] In 1785 Reid could say: "The senses have a double province: they furnish us with sensations, some pleasant others painful and others indifferent. . . ." But by 1905 Freud had cause to remark that "everything relating to the problem of pleasure and pain touches one of the weakest points of present-day psychology,"[39] and that is not far off the truth even today.

The Unicorn Tapestries at the Cluny Museum in Paris, woven in the fifteenth century, depict the five senses, characterizing each sense in terms of the enjoyment it provides: taste—the taste of fruits; smell—the smell of flowers; touch—the touch of a caressing hand; hearing—the sound of music; sight—the sight of beauty reflected in a mirror. But a modern textbook of sensory psychology is unlikely to make more than passing reference to the fact that people may *like* or *dislike* what they feel: that, as Lord Byron wrote, "The great object of life is sensation—to feel that we exist, even though in pain."[40] C. L. Hardin's otherwise excellent survey, *Color for Philosophers*,[41] relegates any mention of the aesthetics of color to a footnote.

Now this bias needs redressing. In fact unless and until we bring sensory affect back into consideration, we shall be fishing for consciousness in an empty pool.

· 5 ·

"WHAT DO WE SEE?"

Vision is *the* dominant human sense; it is the sense that has been most widely studied by psychologists and mulled over by philosophers; and it is the sense for which the distinction between the intimate role of sensation and the defining role of perception is most difficult to draw.

By taking the sense of smell to illustrate his argument Reid might almost be said to have been cheating. In the case of smell, no one needs much convincing that sensations can be pleasing or displeasing. And, with smell, it is relatively easy to recognize that sensation is indeed in a different category to perception. Given that the odor of a rose *enters* my nostrils, my sensation of sweetness is evidently related to "what is happening to me"; while, given that the odor *emanates from* the rose, my perception of the rose as sweet is evidently related to "what is happening out there." Besides, we actually use our noses in two obviously different ways, depending on whether it is subjective feeling or objective definition we are interested in. When we want to savor a smell, we inhale long and deep, but when we want to find out what an object smells of, we typically take a series of short sniffs.

With vision however the situation is never so straightforward. The affective role of visual sensations, though arguably present, is not nearly so striking as with smells. And neither is it so intuitively obvious that visual sensation and visual perception are different categories of experience. I could, it is true, repeat the formula above and say that, given that the light from a rose's petals is falling on my retina, my sensation of redness is evidently related to what is happening to me; while, given that the light is coming from the rose, my perception of the petals as red is evidently related to the external object. But I would not expect the "evidently" here to carry much conviction. Moreover it would be stretching a point to suggest that there are actually two ways of using our eyes, a passively receptive way and an actively exploratory way—for there is certainly nothing quite equivalent to visual savoring as distinct from visual sniffing.

Maybe it is for these reasons that vision has caused so much anxiety to philosophers. Wittgenstein wrote: "We find certain things about seeing puzzling, because we do not find the whole business of seeing puzzling enough."[42] Maurice Bowra, in his *Memories*, tells the story of an Oxford lecturer: "One term he lectured on 'What do we see?' He began hopefully with the idea that we see [subjective] colours, but he abandoned it in the third week, and argued that we see [objectively colored] things. But that would not do either, and by the end of term he admitted ruefully, 'I'm damned if I know what we do see.' "[43] To this philosopher at least it cannot have been obvious that the answer to his question was that vision has a double province, providing us at one and the same time with information about what is happening at our own boundaries *and* information about what is happening in the external world.

Vision therefore provides a special challenge for the kind of account that I am putting forward. It also provides a special opportunity for advancing the argument into new ground.

To begin that advance we must consider how, in evolutionary history, the visual sense began as a *surface sense* whose first role was to provide intimate information about what might almost be called the "smell"—or "taste" or "touch"—of light arriving at the skin.

The most primitive organisms did not of course have eyes (any more than they had noses). Like present-day amoebae they were probably sensitive to light all over their body surfaces. What is more, they did not have specialized "photoreceptors" that were sensitive to light alone: the same sensory receptors might have been responsive not only to light but also to high salt concentration or mechanical vibration.

When photoreceptors did evolve they were not an entirely new kind of receptor. They were simply nonspecific receptors that had evolved to be relatively more sensitive to light than to the other kinds of stimulation. In fact it seems likely that in many cases they developed from "sensory cilia." Cilia are hair-like structures that stick out from the surface of a cell and can serve either in a motor capacity to move the animal around, or in a sensory capacity to detect local disturbances in the environment. By packing a sensory cilium with photosensitive pigment, it could be made to be specifically excitable by light. Even the rods and cones in the retinas of our own eyes show evidence of having started out this way in evolution—as cilia that were sensitive primarily to touch.

The function of photoreceptors in the earliest organisms must have been to detect the general level of illumination. If the light level was "good" the animal could stay where it was, and if it was "bad" it could move about until things improved. But, without any way of telling where the light was coming from, it would have taken a long time to achieve the desired state. And it would not have been until animals developed the ability to compare the local illumination falling on different parts of their body surfaces that they would have been able to move purposively in the right direction.

An earthworm, like an amoeba, has photoreceptors all over its body surface. Earthworms do not like illumination (being at risk from daytime hazards in the open). If a flashlight is shone on a worm on the lawn at night, it rapidly turns away. The worm is comparing what is happening on the bright side of its body with what is happening on the dark side, and on the basis of this comparison it is able to direct its escape. A frog too has photoreceptors throughout its skin (although it additionally has well-formed eyes). By contrast to worms, frogs (being animals better adapted to daylight than to darkness) do like illumination—and they still like it when their eyes are not in use. If a frog with its eyes covered is put in a dark box with a window to one side, it will swing its body around to face the light. Again it is comparing one side with the other.

But is it too soon in evolution, if not in terms of this discussion, to ask "What does an earthworm, or a frog with its eyes closed, *see?*" Given that Bowra's philosopher had such trouble with "What do we see?" perhaps it is a little foolish to start asking the same question about earthworms. But in fact the worm's case may be easier.

Everyone would agree, I think, that the way the worm is representing the light should not be counted as visual perception. But it is at least arguable that it should be counted as visual sensation. For—provided we put aside any worries we may have about whether worms are conscious—it surely makes sense to say that the worm's nervous system is representing the light as "something happening to me," and as something "disagreeable."

For us human beings it is of course hard to imagine how it would feel to be sensitive to light over our entire skin. And yet our own more intimate senses provide a possible way in. If I try to put myself in a worm's place, I can imagine myself being touched, tickled, pained by the light falling on my body; I can imagine the light having a bad taste, or a nasty smell.

But in this case, if the parallel is more with touch or smell or taste than vision, why even suggest that the worm is en route to having a *visual* sensation? I want to do so because, in the history of evolution, the responses of primitive animals to the "touch of light" were in direct line to our own visual experience.

What happened in evolution was that photoreceptors at the body surface became clustered together as "eyespots." Even single-celled animals sometimes have a specialized light-sensitive patch where the threshold for light stimulation is much lower; and most multicelled animals, that do not have proper eyes, have one or more such patches strategically located at their boundaries. The reason for developing these eyespots was that it is more efficient to compare the illumination at several specific locations than to compare the illumination over wide areas of the body.

There proved however to be a better way still of finding out about the direction of a source of light: and this was to transform a single eyespot into a genuine "eye" with an image-forming mechanism (Figure 1). When light from one direction falls on a flat patch of photoreceptors, the patch is evenly illuminated and there is no way of telling which direction the light is coming from; but when the patch is transformed into a cup, light from one direction produces a gradient of illumination; and when the cup is further transformed into a spherical cavity with a narrow aperture at the surface, the arrangement becomes a kind of "pinhole camera," where the direction of the light is precisely correlated with the position of the image. It is only a small step further to fill in the pinhole with a translucent droplet, to produce a full-blown camera with lens.

Camera-like eyes appeared early on in evolution, and have been reinvented several times. But despite their image-forming properties,

Flat Eye Spot

Eye Cup

Pinhole Eye

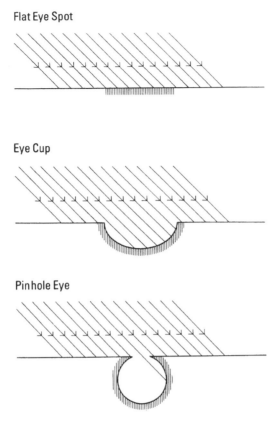

Figure 1

I would suggest that originally their only important function remained that of assessing the level and direction of illumination arriving at the body surface. So, even after eyes had evolved, the sense of vision at first had only a single province, not a double one. When, for example, the image of a bright object moved across the retina, the only experience the animal would have had would have been that of being, as it were, "stroked" by the visual stimulus.

But this is not where evolution rested. Once the image-forming eye had been invented, a whole new world was potentially opened up for *perceptual* analysis. Different-shaped objects, for example, cast different-shaped images on the retina; objects at different distances cast different-sized images; different-colored objects cast different-colored images. Thus light stimulation had become in principle a source of information about the outside world.

By developing a separate channel for visual perception, alongside the already existing channel for visual sensation, animals could take advantage of the defining properties of light, while retaining their primary interest in light as an intimate event affecting their own bodies. The end result, hundreds of millions of years later, was the evolution of animals with eyes and minds like yours or mine: who, when we look at a rose, have the complex and multifaceted experience that we call "seeing."

It could be argued that in our own case the chief function of vision is now perception, and that the affective role of visual sensation has become relatively less important. It is however a general rule of evolution that animals seldom entirely forget their history. Our blood today preserves the same concentration of salts as existed in the oceans from which our ancestors originally emerged. Likewise our experience of seeing preserves, I suggest, memories of the time when light touched us as closely as the scent of a rose entering our nostrils.

But there is another general rule of evolution, and this is that as biological structures or capacities become less important in their original role, new roles get found for them. We might therefore well expect that visual sensation has come to play a secondary role in human mental life for which there is no analogy at all in earthworms.

It would be a mistake, nevertheless, to pass too quickly to thinking about what else, other than affect, the sensation of vision is giving human beings. For, granted that at some crude level visual sensations have less power to move us than do sensations of smell or taste or touch, it is hardly the case that we have evolved to a point where we have ceased to care at all about the light that enters our eyes. We may no longer have photoreceptors all over our bodies. Our retinas, as a proportion of our total skin, may be very small. But then—and I think the point does not need amplifying—a woman's clitoris, as a proportion of her total skin, is very small: and yet clitoral sensations can affect her total being.

· 6 ·

COLOR IS THE
KEYBOARD

In *almost all* circumstances human beings prefer lightness to darkness. Not for nothing does the sun god, the light of the world, outrank every other deity as an object of human worship. Nor for nothing do people feel bright when they are happy and have dark thoughts when they are sad.

When, however, the poet Andrew Marvell wanted to find true comfort, he sought out in his garden "a *green* thought in a *green* shade."[44]

It is not light as such but color that has the most obvious influence on people's moods. Wassily Kandinsky said: "Colour is a power which directly influences the soul. Colour is the keyboard, the eyes are the hammers, the soul is the piano with many strings. The artist is the band which plays, touching one key or another to cause vibration in the soul."[45] But even when no artist is involved and the band is playing a single note, colored light can powerfully affect the human condition.[46]

Red light, for example, has been found to induce physiological symptoms of arousal: blood pressure rises, respiration and heart beat both speed up, and the electrical resistance of the skin decreases. By contrast, blue light has the opposite effect: blood pressure falls slightly, and heart rate and breathing slow down. These responses are almost certainly unlearned. At only fifteen days old, crying infants can be more readily quieted by blue light than by red.

People feel subjectively warmer in red rooms than in blue. W. E. Miles reported that in a café the women employees found that they could discard their coats when the blue walls were repainted orange.

A Norwegian study showed that people would set a thermostat four degrees higher in a blue room than they would in a red one, as if to attempt a thermal compensation for the coolness that was visually induced.

Subjective time may pass faster in red light than in blue, so that people judge, say, a minute in a red room to be equivalent to one and a half minutes in a blue one. The reaction times of a group of students were reported to be faster when the room illumination was red than when it was green. A study in a factory showed that workers spent less time in the bathrooms when they were painted a deep red.

In a book on *Colour for Architecture*[47] Tom Porter and Byron Mikellides relate that "Michelangelo Antonioni, the Italian film director, made an interesting observation during the making of his first colour film, *The Red Desert*. While shooting industrial scenes on location in a factory, he painted the canteen red in order to invoke a mood required as background to the dialogue. Two weeks later he observed that the factory workers had become aggressive and had begun to fight amongst themselves. When the filming was completed the canteen was repainted in a pale green in order to restore peace and so that, as Antonioni commented, "The workers' eyes could have a rest.' "

Furthermore, "Clinicians and art therapists have observed that suicidally-inclined patients tend to use yellow pigment generously in their paintings—as, indeed, did Vincent Van Gogh. His last painting before committing suicide was the mainly yellow *Wheatfield with Crows*. . . . The Institute of Contemporary Arts in London discovered to their cost that the stimulant effect of yellow is so intense that it can incite children to vandalism. During an exhibition of toys, displayed in various coloured rooms, all those in the yellow room were damaged or broken!"

In certain pathological conditions, the effects of color can become still more pronounced. Kurt Goldstein described a patient with cerebellar disease: "If she was clad in a red dress, all her symptoms increased to an unbearable degree. She became dizzy and fell. Green or blue had the opposite effect. They made her quiet; her equilibrium improved so that she appeared to be almost normal."[48] He further observed that, with this woman and other patients with cerebellar damage, looking at a red or yellow screen caused the arms to swing away from the body, while a green or blue one caused them to come closer. L. Halpern described several similar cases. In one of them: "When a pure red glass was placed before the patient's left eye, her whole body started rocking immediately . . . while at the same time her right arm descended and deviated widely to the right. . . . The patient stated that, when looking at red, breathing became difficult and palpitation and nausea developed. In contrast to these disturbing sensations . . . the patient felt subjectively completely well when using a blue glass."[49] In red light, pain sensitivity increased, and loud noises that would be tolerated in blue light became unbearably unpleasant.

Goldstein concludes: "The stronger deviation of the arms in red stimulation corresponds to the experience of being disrupted, thrown out, abnormally attracted to the outer world. It is only another expression of the patient's feelings of obtrusion, aggression, excitation, by red. The diminution of the deviation in green illumination corresponds to the withdrawal from the outer world and retreat into his own quietness, his centre."

These muscular reactions may be observable in a more muted form even in healthy people. The musician Manfred Clynes developed a technique for measuring emotion, using a sensitive pressure pad—the "sentograph"—which picked up tiny expressive movements in a subject's finger. In his book *Sentics: The Touch of Emotions* Clynes shows a typical response to red as a "strong outer-directed response," while "the calm of blue is reflected by . . . the absence of an outward thrust."[50] The forms of the sentograms to red and blue are remarkably similar to those he found when he asked subjects to contemplate the states respectively of hate and friendship.

Although it must be said that much of this research on color is relatively second-rate—reflecting the general modern bias in psychology against studying affect—the overall picture that emerges is of human beings as animals who have retained a strong biological memory of light as an intimate event. We may indeed not be so different from our long-distant ancestors, who sensed light with their whole skin—and let the vibrations get through, if not to their soul, to their muscles and their glands.

IN THE REALM
OF THE SENSES

For *Samuel Coleridge* the experience of vision had connotations that were transparently erotic: "Sometimes when I earnestly look at a beautiful object or landscape, it seems as if I were on the brink of a fruition still denied—as if Vision were an appetite; even as a man would feel who, having put forth all his muscular strength in an act of prosilience, is at the very moment held back—he leaps and yet moves not from his place."[51]

William Wordsworth, recalling his youth, described himself in love with form and color:

> . . . *the tall rock,*
> *The mountain, and the deep and gloomy wood,*
> *Their colours and their forms, were then to me*
> *An appetite; a feeling and a love,*
> *That had no need of a remoter charm,*
> *By thought supplied, nor any interest*
> *Unborrowed from the eye.*[52]

Writing in the 1790s, in the climate of ideas created by Reid, Wordsworth well understood the distinction between sensation and perception. It was not perception that he craved for, not the "remoter charm by thought supplied," but rather the raw sensation of light, containing nothing "unborrowed from the eye."

.   .   .

Instead of saying, as I did, that the intimate role of visual sensation is nothing like so obvious as it is with smell, perhaps I should have said that the defining role of visual perception is much more obvious: and that it is because vision is such a remarkable source of objective external information that the similarity between vision and the lower senses has often been discounted.

Plato distinguished sharply between the "higher" senses of vision and hearing and the "lower" senses of smell, taste, and touch, exalting the former only as pathways to rational knowledge: "God devised the gift of sight for us so that we might observe the movements which have been described by reason in the heavens, and apply them to the motions of our own mind. . . . And the same holds good of voice and hearing."[53] He was aware that vision and hearing could, like the other senses, also excite what he called "irrational pleasure" at the level of mere sensation. But to let oneself be ruled by sensation ought to be morally repugnant to people of good taste and virtue.

When classical Greek ideas reached Europe at the beginning of the Renaissance, this Platonic prejudice was reinvoked. Boccaccio for example could write that Giotto "had brought back to light that art that had been buried for centuries under the errors of those who painted rather to delight the eyes of the ignorant than to please the intellect of the wise."[54]

Two hundred years later the Unicorn Tapestries at Cluny depicting the five senses struck the same moral stance. The first five tableaux, as I mentioned earlier, celebrate sensory pleasure. But I did not mention that the sixth shows the Lady with the Unicorn putting a string of jewels back into a casket, and on the awning over the tent is written "A mon seul désir"—"At my sole behest." She is saying, like a good Platonist, that she will renounce the seductive pleasures of sensation in order not to cloud her rational mind.[55]

Yet, as the Roman poet Horace wrote, you can drive out nature with a pitchfork, and she will always return. In art and poetry, people's delight in sensation simply went underground, to return with new champions in the eighteenth and nineteenth centuries. Wordsworth, speaking for the English Romantic movement, was entirely scornful of those who would disparage the intimate enjoyment of the senses.

Up! up! my Friend, and clear your looks;
Why all this toil and trouble?

Up! up! my Friend, and quit your books;
Or surely you'll grow double:

One impulse from a vernal wood
May teach you more of man,
Of moral evil and of good,
Than all the sages can.

. . .

The eye—it cannot choose but see;
We cannot bid the ear be still;
Our bodies feel, where'er they be,
Against or with our will.[56]

In England the painter William Turner, and later in France the Impressionists, took up the invitation to paint "nothing unborrowed from the eye" and set out deliberately to feed the appetite for visual sensation by creating paintings that not only made no concessions to perception but often actually worked against it. In Turner's later landscapes, for example, the artist made light itself the subject of his painting, representing the wash of colors arriving at his retina as exaggerated brushstrokes on the canvas. Land, sea, ships, cattle had lost all definition—so that what we now experience when we look at his paintings is not the image of external objects but simply the caress of light.

In the same spirit Claude Monet could paint more than twenty different pictures of Rouen Cathedral, seen from more or less the same angle but under different conditions of light and weather. The object of perception in every picture remained constant (the ability to "see through" the vagaries of stimulation is one of the major achievements of perception); but in each case the sensation was wonderfully different.

John Constable accused Turner of painting "in tinted steam";[57] someone else said that his landscapes were "pictures of nothing and very like." But Turner and Monet might almost have been following Immanuel Kant's suggestion, in the *Critique of Judgement*: "When the question is whether a thing be beautiful, we do not want to know whether anything depends on the existence of the object, but only how we estimate it in mere contemplation."[58] By deliberately suppressing the existence of the object they were assisting the viewer to a state of visual contemplation.

. . .

Paul Cézanne believed that human beings who are too concerned with the "existence of the object" may miss out on sensation altogether. Of a farmer who drove him to market, he remarked: "He had never seen, what we would call seeing; he had never seen Sainte Victoire. He knows what has been planted there, along the road, how the weather is going to be tomorrow, whether Sainte Victoire has his cloud cap on or not; . . . but that the trees are green, and that this green is a tree, that this earth is red, and that this red rubble and boulders are hills, I really do not believe that he feels that."[59]

Just as a wine taster may temporarily put to one side his pleasure in the gustatory stimulus in order to focus on the question of what the wine is made of, so someone may not notice the beauty of light when what concerns him is entirely what lies out there in the material world.

But many of us much of the time are—for the best of biological reasons—in the same condition as the farmer. And to "cleanse the doors of perception," as William Blake put it, requires a kind of disinvolvement with reality that does not come easily. Wordsworth recommended quiet passiveness. Others, especially religious mystics, have used contemplative exercises. But quicker and probably more effective (and more certain to upset the rationalists) is the use of psychedelic drugs.

Aldous Huxley described his own experiment with mescaline: "Visual impressions are greatly intensified and the eye recovers some of the perceptual innocence of childhood, when the sensum was not immediately and automatically subordinated to the concept. . . . The books, for example, with which my study walls were lined. Like the flowers, they glowed, when I looked at them, with brighter colours, a profounder significance. Red books, like rubies; emerald books; books bound in white jade; books of agate; of aquamarine, of yellow topaz. . . . At ordinary times the eye concerns itself with such problems as Where?—How far?—How situated in relation to what? In the mescaline experience the implied questions to which the eye responds are of another order. Place and distance cease to be of much interest. The mind does its perceiving in terms of intensity of existence. . . . Until this morning I had known contemplation only in its humbler, its more ordinary forms. . . . But now I knew contemplation at its height."[60]

Lest the report of someone in this state of ecstasy should not be trusted, here is a comparable description from a woman who had taken LSD: "About three-quarters of an hour after the beginning of the experiment a different quality of consciousness came with a rush. Nothing was definably changed, but the room was suddenly transfigured. All objects stood out in space in an amazing way and seemed luminous. I was aware of the space between objects, which was pure vibrating crystal. Everything was beautiful. . . . I said 'It is poignantly lovely, but I can't explain why. There is a divine ordinariness about it and yet it is completely different.' "[61]

What both are describing is the intensification of visual sensation and the overwhelming of perception: paradoxically, a semimystical experience achieved when the God-given faculty of outward-oriented sight is pushed—by chemicals—to second place.

· 8 ·

SHUTTLE VISION

magine someone writing with a feather on the skin of your back, and compare what it would be like to relish the tactile stimulus as against perceiving what is written. Imagine hearing the Moonlight Sonata, and compare basking in the music as against trying to discern whether the pianist is Richter or Serkin. Ask a professional wine taster whether he has actually enjoyed the claret he has just recognized as a 1970 Lafite, and he very likely will not be able to say.

Perception and sensation do indeed involve different kinds of attention or different attitudes of mind. Some years ago I conducted a series of experiments with rhesus monkeys which provided an unexpected demonstration of how they can shift in and out of attending to sensation and perception.[62]

The experiments were concerned in the first place with investigating the monkeys' affective response to colored light. I put each monkey in a dark testing chamber with a screen at one end onto which one of two alternative slides could be projected. The monkey could control the presentation of the slides by pressing a button, each press producing one or other slide in strict alternation. Thus when he liked what he saw he could hold the button down, but if he wanted to change he could release and press again.

To test for "color preference" I let the monkeys choose between two featureless fields of colored light of the same brightness. The result was that all eight monkeys that were tested showed strong and consistent preferences. When given a choice between, for instance, red and blue, they spent three or four times as long with the blue as the red. Across the spectrum the order of preference was blue, green, yellow, orange, red. When each of the colors was separately paired with a neutral white field, red and orange stood out as strongly aversive, blue and green as mildly attractive.

In a separate experiment, instead of giving the monkeys a button to change the light, I let them move bodily backward and forward between two chambers that were permanently lit.[63] Again they preferred a blue chamber to a red one. And if *both* were red, they shuttled rapidly backward and forward, as if highly uneasy; while if both were blue they settled down. Their dislike for the red light became all the greater when their choice was made in the presence of loud and unpleasant background noise.[64] All in all, these monkeys were showing rather the same reactions as human patients with cerebellar disease.

Now, in the context of the preceding discussion, the question that might be asked is this: Were the monkeys' preferences being determined by sensation or perception? Was it the subjective experience of being bathed in red light that they hated or was it the objective fact that everything was colored red? Since there was nothing obvious for the monkeys to look at in the chamber, and hence very little to engage their perceptual faculties, it seemed highly probable from the start that it was the sensation of redness that affected them. But what really persuaded me that this was so was what occurred when there *was* something for the monkeys to look at.

In the situation where they could change the slide by pressing a button, I first gave them a choice between the white field and an "interesting" black-and-white motion picture showing Mickey Mouse. Monkeys are inquisitive animals, and not surprisingly they showed a strong preference for the picture. But then I projected the picture through a red filter so that it became a black-and-red picture where everything in it was colored red. You might have guessed that the two factors—their interest in the pictorial content and their dislike for the redness—would have canceled each other out. But no, the result was that now the red light had no effect at all: the monkeys were just as keen to watch the picture as if it had been black-and-white.

To put some numbers on this, the results of a particular experiment with two monkeys were as follows. When the choice was between featureless red and white fields, they chose the red field 29 and 28 percent of the time. When it was between a black-and-white moving picture and a white field, they chose the picture 84 and 86 percent of the time. When it was between a black-and-red picture and a white field they still chose the picture 83 and 86 percent of the time.

In further tests I used repetitive brief film loops, so that the monkeys would eventually find nothing new to look at. I found that as and when their interest in the pictorial content evaporated, then they reverted to strongly preferring the white field. Mathematical analysis of these and other results showed that the monkeys' behavior could be closely fitted by a two-factor theory where "perceptual interest" and "sensory pleasure/unpleasure" were assumed to be completely independent variables, and the former overrode the latter.

It was as if monkeys, like people, can be attentive either to perception or sensation, but not easily to both. Like Cézanne's farmer or the wine taster, when they switched to a perceptual mode—an allocentric or defining mode—their interest in the existence of the external object was dominant; but when they switched back to a sensory mode—an autocentric or intimate mode—their feelings about the color of the light came through.

Roger Fry, the painter and critic, noted a very similar double experience in people's response to paintings.[65] Many great paintings, according to Fry, appeal to us both at the "dramatic or psychological level"—by which he meant their pictorial, storytelling content— and at the "plastic" level—by which he meant their aesthetic content determined simply by the arrangement of color and form. But these two are frequently in competition, so that "we are compelled to focus the two elements separately. . . . What in fact happens is that we constantly shift our attention backwards and forwards from one to the other"; but as a work becomes familiar the "psychological elements will, as it were, fade into the second place, and the plastic quality will appear almost alone."

I said earlier that in the case of vision there is no obvious equivalent to savoring as distinct from sniffing. But actually, both for people and monkeys, it seems there is. What is more, in the human case, which way we "use our eyes" is at least to some extent under our voluntary control. Particular sights and circumstances may push us one way or the other, but even then we can go against the bias if we please. When we stand before Monet's painting of Rouen Cathedral, we can if we wish reject the invitation to bask in the visual stimulus and concentrate instead on what we can make out of the external object; but equally, when we stand before the real Rouen Cathedral, we can (precisely because Monet has helped us do it) reject the call of the external object and concentrate instead on the stimulus arriving at our eyes.

But I must be careful with my choice of examples or I shall create the false impression that I am not talking about ordinary experience. The truth is we can and do see everything in these two ways. What is true of the cathedral is just as true of the yellow pencil on my desk. I can represent it as a pencil or as a streak of light arriving at my retina (and if I bring it too close to my eyes I find myself experiencing the retinal stimulation twice over, while never doubting that there is only one objective pencil).

It takes some practice to switch visual modes at will. It is not always easy, as Reid said, "to attribute nothing to one which belongs to the other." But it can be done: which is just as well, because the argument of the coming chapters will rely on it.

· 9 ·

"IT MUST LOOK
QUEER!"

The next few chapters deal with relatively technical issues, and before embarking on them I ought to explain why it is appropriate to worry away at problems that might be thought more suited to a textbook on sensory psychology.

John Locke wrote in his *Essay Concerning Human Understanding:* "Let any one examine his own thoughts, and thoroughly search into his understanding, and then let him tell me, whether all the original ideas he has there, are any other than of the objects of his senses, or of the operations of his mind considered as objects of his reflection."[66]

The senses, as Locke recognized, are almost literally the gates and windows to the mind, through which all new information passes; so that there can be no thoughts, ideas, conceptions in our head that do not derive originally from our experience of surface stimuli impinging on our bodies. But the question of exactly how people or animals interpret surface stimuli—how they deal with information at the boundary between "me" and "not-me"—has been and remains surprisingly contentious.

Are sensation and perception really different, and if so how? When I now look at a patch of color, or smell a rose, or feel pain, are there really, as Reid (and I) maintain, two things going on or only one? And if we can answer the question in our own case, then what about other animals? What is it like to be a bat, echo-locating its way through space? Or a pigeon navigating by a magnetic sense? Or for that matter a mechanical robot with artificial sense organs and an electronic computer for a brain? Are there animals or machines that

have sensation and no perception . . . or perception and no sensation
. . . or perception with different sensation? And if any of these
actually occur, how would we know? The questions lead on directly
to the apparent privacy of individual experience, and the famous
problem of Other Minds. Are my pains like yours? How do I know
that you feel pain at all?

If the fish of *consciousness* is lurking somewhere, it is surely in this
area of the river. But the reason it has not been caught is partly at
least that theorists have been overquick in assuming they know *a
priori* what sensory experience amounts to. As Bertrand Russell
wryly noted in the *Introduction to Mathematical Philosophy*: "The
method of 'postulating' what we want has many advantages; they
are the same as the advantages of theft over honest toil."[67]

A notorious "thought experiment" can be used to illustrate what
is at stake.

THE "INVERTED SPECTRUM"

Imagine a color negative in which greens are red, blues are yellow,
and so on—grass looks the color of blood, ripe tomatoes look like
unripe ones, and marigolds look the color of violets. Suppose there
were spectacles you could wear which produced a "color spectrum
inversion" in the light arriving at your eye, so that the colors of the
retinal image were transposed in just this way. What would be the
short-term and the long-term consequences of wearing spectacles
like these?

Provided one accepts the distinction between sensation and perception, it is obvious what must happen. When you first put on the
spectacles both your sensation and your perception would be altered: you would have the sensation of green when you looked at a
ripe tomato, and equally you would perceive the surface color of the
tomato to be green—so that you would call it "green" and might
actually mistake it for an unripe one. Indeed if, like the poet, you
wanted "a green thought in a green shade" you might now choose
to sit in a red room instead of a green garden.

In the longer term, however, your experience would presumably
change. There is no reason to suppose that your sensation would
ever go back to what it was, since when red light fell on your

spectacles the light at your retina would be not red but green, and your assessment that something green was happening to you would always remain valid. On the other hand there would be every reason to suppose your perception would eventually get back to normal, since whenever you mistook the colors of external objects you would be liable to be corrected by events. Thus, while your sensation remained transformed, your language and objective judgments about colored objects would probably revert soon enough to what they were before. Note, however, that, if *affective* response is determined chiefly by sensation, you would still prefer a red room to a green garden—only now you might say that you were seeking "a red thought in a red shade."

The color-inversion experiment has never been done, and practical limitations probably mean it never will be. But thought-experiment versions of it have been widely discussed by philosophers. Locke started it by considering the possibility, not that a single individual might don color-inverting spectacles, but that separate individuals might differ from birth in the structure of their eyes, so that while always having different sensations of color they grew up to make correct perceptual judgments:

"If by the different structure of our organs, it were so ordered, that the same object should produce in several men's minds different ideas at the same time; e.g. if the idea, that a violet produces in one man's mind by his eyes, were the same that a marigold produced in another man's, and vice versa . . . he would be able as regularly to distinguish things for his use by those appearances, and understand, and signify those distinctions, marked by the names blue and yellow, as if the appearances, or ideas in his mind, received from those two flowers, were exactly the same, with the ideas in other men's minds."[68]

By taking the case of different individuals rather than a single individual who undergoes a change, Locke could raise the tantalizing possibility that "this could never be known: because one man's mind could not pass into another man's body, to perceive, what appearances were produced by those organs."

Indeed, ever since Locke, philosophers have gone around wondering out loud whether it might actually be the case that different members of the human species really do experience colors differently without anyone knowing it. Wittgenstein wrote in the *Philosophical Investigations*: "The assumption would thus be possible—though unverifiable—that one section of mankind has one sensation of red and another section another."[69]

But could that really be so, that "this could never be known" and that it would be "unverifiable"? Only, of course, if it were to be true that sensations of colored light make no difference to the way a person behaves. And I have been arguing in the preceding chapters that just the opposite is true: that sensations matter, and that in particular there is almost certainly a nonarbitrary connection between sensation and affect.

Wittgenstein himself at an earlier stage in his career also raised the possibility that affective responses would bring the truth to light. Here he is considering a case where a single individual wakes up to find his color experience changed (just as if he had been fitted during the night with color-inverting spectacles without his realizing it): "Consider this case: someone says 'I can't understand it, I see everything red blue today and vice versa.' We answer 'it must look queer!' He says it does and, e.g., goes on to say how cold the glowing coal looks and how warm the clear (blue) sky. I think we should under these or similar circumstances be inclined to say that he saw red what we saw blue."[70]

Now, if what I have been proposing holds, the man would almost certainly go on making these anomalous judgments about the warmth of blue light and the coldness of red light, even after he had reverted to using the correct color names. So that, in his case at least, the assumption that he was having different sensations would never become unverifiable to an outside observer, even if he himself were to forget what his experience used to be. In the case of someone who was born with "color-inverted eyes," I cannot see why exactly the same considerations would not apply.

Admittedly this still skirts around the question of the conscious quality of a person's experience. And no necessary link has been made yet between having a sensation with a particular affective tone and having a sensation with a particular what-it's-like-to-have-it conscious feel. I believe there is such a link: indeed that having sensations that we *mind about* is part and parcel of having experiences that we are *conscious of*. But first it is essential to establish the prior case that sensation is worth taking seriously at all.

And for that we must leave thought experiments and return to the real world. Denis Diderot wrote: "Unfortunately it is easier and quicker to consult oneself than to consult nature. . . . We should distinguish two kinds of philosophy, the experimental and that based on reasoning. . . . The philosophy based on reasoning makes a pronouncement and stops short. It boldly said: 'light cannot be decomposed': experimental science heard, and held its tongue in its presence for whole centuries; then suddenly it produced the prism, and said, 'light can be decomposed.' "[71]

There are in fact philosophers who would bet their bottom dollar that sensory experience cannot be decomposed into sensation and perception, others who have said that obviously it can be. What is needed to help resolve the issue is the equivalent of the experimental prism.

· 10 ·

NEW ARRANGEMENTS

"You are old, Father William," the young man said,
  "And your hair has become very white;
And yet you incessantly stand on your head—
  Do you think, at your age, it is right?"

"In my youth," Father William replied to his son,
  "I feared it might injure the brain;
But now that I'm perfectly sure I have none,
  Why, I do it again and again."[72]

Lewis Carroll in *Alice in Wonderland* was making fun of Robert Southey, who in his own poem about Father William had made fun of William Wordsworth, the doyen of sensation. Had he but known it, he was also hinting at an important experiment in sensory rearrangement.

As a reference point for the ensuing discussion, let me bring back my diagram showing how sensation and perception are presumptively related, adapted now specifically for vision.

                                        sensation of what is
                                        happening at the eye
external object → light at the retina ⤙
                                        perception of what is
                                        happening out there

UPSIDE-DOWN VISION

Try looking at the world with your head between your legs. If you attend to visual sensation it will be apparent that the image at your retina has now turned around: parts of the image that previously appeared nearer the top of your eye socket are now nearer the bottom, parts that were nearer the right side are now nearer the left, and so on. If however (as probably comes more naturally) you attend to perception, you will find that everything about the external world is still much as it was: the ceiling is still perceived to be above the floor, the text of a book still reads from left to right, and so on. You can easily check on the continuing accuracy of your perception by trying to point to things in the environment: you will find you have no trouble—although note that when you point to an object whose image appears nearer the top of your eye you are now pointing in a different direction than heretofore.

There is nothing surprising or controversial about this. What it shows is that, while you rely simply on the retinal image in order to form a representation of "what is happening at the eye," you are able to—and indeed must—take additional account of the orientation of your head in space in order to form a perceptual representation of "what is happening out there." But it still illustrates an important fact, namely that different visual sensations (a right-way-up or an upside-down image) can indeed be associated with the same perception (an upright world)—provided the perceptual mechanism in your brain is informed of what the situation is, so that it can make the requisite adjustments.

Suppose however that a change were to occur in the orientation of the retinal image without any change in the orientation of your head and hence *without* the perceptual mechanism in your brain being so informed. In particular, suppose you were to wear special "up-down inverting spectacles" in front of your eyes, so that even while you yourself remained upright your retinal image were to be permanently turned the wrong way up. In this situation the perceptual mechanism would make no allowance for the transformation of the image, and so—initially at least—you would see both the image to be the wrong way up (which it is) *and* the external world to be the wrong way up (which it is not). Hence you would be bound to make perceptual mistakes—pointing up for an object when you should point down, calling "top" "bottom," and so on.

What would be the effects of wearing these spectacles for a long time? The case is analogous, at least in principle, to the thought experiment with color inversion. There is no reason to suppose that your sensation would ever get back to what it was, since your assessment of the image as having been turned around on your retina would remain perfectly correct. On the other hand, there would be every reason to suppose that your perception would eventually undergo some sort of readjustment, since whenever you pointed in the wrong direction you would be pulled up short. Thus we might expect that the perceptual mechanism would in fact become "recalibrated" to take account of the new situation, so that it once again gave you a valid picture of the position of objects in space.

This experiment with inverting spectacles has in fact been tried out in practice several times over the last hundred years, with subjects wearing the spectacles for as much as a month continuously. Given the methodological problems of getting people to go about their daily business in a topsy-turvy world, it is hardly surprising that the results of different studies have not been entirely in agreement. There have been problems too in interpreting people's introspective reports, when they say for example that the way "things look" to them (perception or sensation?) has or has not changed.

Nonetheless, in a series of studies conducted at Innsbruck in the 1960s, I. Kohler found clear evidence that after a subject has worn the inverting spectacles for as little as two weeks there can indeed be virtually complete perceptual readjustment: to the point where, for example, the wearer of the spectacles can ride a bicycle or catch a ball and generally relates to the external world as if he were once again perceiving it the right way up. When the spectacles are removed, the subject now makes errors of pointing in the contrary direction. In one experiment Kohler used half glasses, such that the image was inverted when the wearer looked upward but normal when he looked downward, and it turned out that the subject could eventually adapt to that as well, i.e. he could learn to make allowances for the direction of his gaze.[73]

But if perception adapts, what happens to sensation? The results of Kohler's studies and others have been reviewed by Robert Welch in a book on *Perceptual Modification*[4] where he has attempted to distinguish carefully between changes occurring at what he calls the "egocentric" (sensation) level and the "environmental" (perception) level. Welch concludes that even when perceptual readjustment is complete, there is apparently no corresponding adjustment in sensation: "critical introspection," as he puts it, shows that the retinal image still continues to appear to be the wrong way up compared to what it was before. Correspondingly, after subjects remove the spectacles, even though they make perceptual errors they report that their sensory experience has returned to being "familiar."

There seems to be no doubt therefore that the predicted dissociation between sensation and perception can occur, not just in thought experiments but in real life. The next example makes the point even more forcefully.

SKIN-VISION

Given that the human retina began life in evolution as a part of the skin, perhaps it could be said that what we all have already is skin-vision (and by the same token skin-taste, skin-smell, and skin-hearing). Earlier on I was pressing the metaphor of human beings and other animals responding to the "touch of light." And yet there is clearly "skin" and "skin": skin that has been transformed into a light-sensitive retina and plain old skin. Common sense would suggest that no one could ever *see* with the skin of his back.

There are two obvious problems: first that the skin of the human back lacks light receptors, and second that even if a person did have light receptors he would still lack any kind of image-forming mechanism—so that all he could detect would be the general level of illumination. Suppose however that both these problems could be got around. Suppose that an artificial lens were used to form the light into an image and then this image were transformed into a form of stimulus to which the skin is sensitive such as vibration. Is it not possible that the information arriving at the skin might then be quite adequate to provide—with extended practice—a basis for recognizing what the light signified in the external world? Moreover that this would work just as well for blind people as for normally sighted ones?

In the late 1960s Paul Bach-y-Rita and his colleagues at the Smith Kettlewell Institute undertook some trials with a "sensory substitution apparatus" based on just this reasoning.[5] What they did was to provide the subject with a tiny TV camera attached to his head, whose electronic image, instead of going to a TV screen, was sent to a matrix of vibrators in contact with the skin of the back. There were 400 vibrators in a 20 × 20 matrix, covering a 10-inch-square area of skin. Thus each point stimulated on the skin represented one small area of the image captured by the camera, much as a newspaper photograph represents a scene by an array of dots. The subject could direct the camera by moving his head, rather as if he were moving his own eyes.

The results surpassed all expectations. With only a few hours' training, blind subjects learned to recognize a range of common objects such as a telephone, a cup, and a toy horse. Very quickly they developed the ability to point accurately to objects in space, and to judge their distance and their absolute size (independent of distance). After about thirty hours of training they could make complex pattern discriminations and some subjects even learned to recognize the faces of members of the laboratory staff. Bach-y-Rita quotes an experienced subject engaged in exploring the visual scene with the camera: "That is Betty; she is wearing her hair down today and does not have her glasses on; her mouth is open, and she is moving her right hand from her left side to the back of her head."

Perhaps most remarkable of all was the evidence for spatial perception. By making use of information in the image about perspective and parallax, the blind subjects came to perceive external objects as being located in a stable three-dimensional world. They did not locate objects as lying up against their skin—any more than we with normal vision locate objects as lying up against the retina of our eyes—but immediately perceived them as being out there in space.

Bach-y-Rita has no compunction about saying that his blind subjects acquired *visual* perception: "If a subject without functioning eyes can perceive detailed information in space, correctly localize it subjectively, and respond to it in a manner comparable to the response of a normally sighted person, I feel justified in applying the term 'vision.' "

I would agree with him. But what about sensation? True to the bias of modern psychology, Bach-y-Rita actually has very little to say about sensation. The question is nonetheless an obvious and an interesting one: when a blind person *sees* with the skin of his back, does he experience *visual* or *tactile* sensations? Presumably in the very first minutes of trying out the apparatus he must have tactile sensations—sensations of himself being touched—for there is no reason yet why his experience should be any different from yours or mine. But as he learns to interpret the tactile stimulus as a visual percept, it is I suppose conceivable that he would begin to have sensations as if of light arriving at his retina, in other words *visual* sensations of light and dark.

I know one clever philosopher whose first pass at trying to imagine himself in the blind subject's place was to say, "Yes, of course his sensations would be visual." But this is surely counterintuitive. Whatever the subject makes of the stimulus perceptually, the fact remains that he is not being stimulated by light at his retina, he is being stimulated by mechanical vibration on the skin of his back. And insofar as sensation is a representation of "what is happening to me," there is no reason whatever why its quality should change when "what is happening to me" continues to be what was originally felt as tactile stimulation.

There is however another possibility, which is that the subject might have no sensations at all. For he might be so absorbed in the task of perceiving the external world that he would switch over entirely to the perceptual mode and discount his sensation altogether.

But I am discussing this as if it were a thought experiment, when the fact is we ought to have the evidence of real live people. And although Bach-y-Rita is relatively silent on this question he is not completely silent. In his book on *Sensory Substitution* he writes: "Even during task performance . . . the subject can perceive purely tactile sensations when he is asked to concentrate on those sensations." However, "unless specifically asked, experienced subjects are not attending to the sensation of stimulation on the skin of their back, although this can be recalled and experienced in retrospect."

Thus it seems that most of the time the subject is indeed simply unaware of what is happening to him; but if and when he reminds himself of what it feels like at the level of sensation, his experience is unambiguously tactile.

In summary, we have these two contrasting situations: the case of normal vision, and the case of skin-vision:

Normal vision

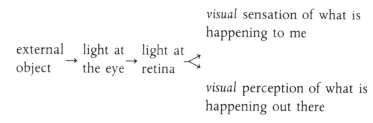

Skin-vision

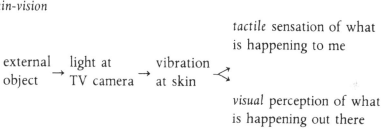

The plot is thickening. It gets thicker still when one considers what might happen if there were to be a selective breakdown in either sensation or perception.

· 11 ·

MIND-BLINDNESS AND
BLIND-MINDNESS

A*lice in Wonderland,* having
gone down the rabbit hole
and drunk the liquid in the bottle labeled DRINK ME and eaten the
cake in the box labeled EAT ME, began to experience a variety of
strange symptoms. One moment she seemed to shrink in size, the
next she opened out like a telescope. She found a golden key, and
with it she opened the door to a garden where nothing was quite as
it seemed. There was a Cheshire cat there, who disappeared leaving
behind only its grin. '' 'Well! I've often seen a cat without a grin,'
thought Alice; 'but a grin without a cat! It's the most curious thing
I ever saw in all my life!' ''[76]

Lewis Carroll, I can only suppose, was anticipating my argument
again and hinting at the possibility of a pathological dissociation
between sensation and perception. A grin without a cat—perception
without sensation?—would be a very curious phenomenon indeed.
But first let me consider the cat without the grin.

Bad Perception/Good Sensation

We have already had ample evidence of how perception can give the
wrong answers even though sensation gives the right ones. When
someone first puts on inverting spectacles his perception is grossly
in error (he sees the external world upside down), and when someone
first tries out the skin-vision apparatus his perception is absent
altogether (he does not yet perceive the world at all), while in neither

case is there anything wrong with his sensation. In each case perception has to be modified by learning. But if perception can be acquired or altered by experience, it must be all too likely that it can also be impaired by brain disease.

"Mind-blindness" or "visual agnosia" is in fact a well-documented consequence of damage to the association cortex of the brain. ("Agnosia," a term coined by Freud, is literally "not-knowing," but has come to mean specifically the loss of some aspect of perception while sensation is relatively unaffected.)

Typical is a case described by Macdonald Critchley: "A sixty year old man woke from a sleep unable to find his clothes, though they lay ready for him close by. As soon as his wife put the garments into his hands, he recognized them, dressed himself correctly and went out. In the streets he found he could not recognise people—not even his own daughter. He could see things, but not tell what they were. . . . Psychologically, he was completely clear and normally oriented. Intelligence was rather above average." In this patient "there was no disturbance of mentation and the traditional sense-physiological examination revealed no abnormality"; nonetheless, "of large objects, he recognized only a bottle of wine." What had happened was that during the night he had suffered a minor stroke that had damaged his parietal cortex. As a consequence his higher perceptual faculties were disabled while his sensation was left almost undisturbed."

In this case the agnosia extended to many aspects of perception. But in others the agnosia has proved to be remarkably specific. Patients have been described who are unable to perceive shape, or movement, or spatial location, or color; or unable to recognize particular classes of objects, such as faces, or vegetables, or musical instruments. But all the time they will say that their sensation is quite normal—and that nothing looks any different to the way it did before.

"Color agnosia" is a specific difficulty with recognizing the colors of external objects. I examined a case of this kind in Oxford some years ago.[8] The woman patient thought she saw colors as she always had done. When tested for color blindness with the plates that show a colored figure against another colored background, she proved to have normal color sensitivity, and she could quite well sort colored discs into same-color piles. Moreover when asked "what color is a banana?" ". . . a mailbox?" and so on, she was right every time. However, when she was shown pieces of colored paper and asked to say what color she saw them to be she made bizarre mistakes: when shown a piece of blue paper—"red"; green paper—"between red and orange"; yellow paper—"blue." Yet, to repeat, she said the quality of her color vision was quite unaltered—and indeed she was constantly surprised at our taking any interest in this aspect of her case.

What is it like to be agnosic? Anyone who has listened to someone speaking a foreign language and not understood what the sounds signify knows, I think, something of what it is like to have an "auditory agnosia." Most of us have experienced at least a fleeting "visual object agnosia" when we have looked at puzzle pictures and been unable at first to make them out; or a "visual depth agnosia" when we have looked into a stereoscope and at first not seen the 3-D scene.

When someone expects to understand something and finds himself unable to, he is likely of course to be baffled and annoyed. But beyond that, interestingly enough, the patients themselves do not reckon that their experience is all that peculiar. And the truth surely is that it is not *all* that peculiar. As far as the patient is concerned he can still "see," only not see very well; and in fact it is not uncommon for the patient to believe that his only problem is that he needs a change of spectacles.

The agnosias are fascinating in themselves, and of great interest to psychologists concerned with perceptual mechanisms. But I would stress that it would be a mistake to suppose that the patients' experience is totally different to anything we ourselves already know. I say this because I now want to contrast it with the obverse of agnosia where sensation fails while perception stays intact.

BAD SENSATION/GOOD PERCEPTION

If the scheme of two parallel channels is anything like right, the possibility of perception continuing in the absence of sensation is clearly on the cards. But, unlike the agnosias, it is a condition for which most of us have no obvious model in our own experience. Imagine listening to someone speaking and discovering that you understood his meaning but were unaware of any sounds arriving at your ear, or looking at a picture and seeing what it represented but being unaware of receiving any visual image at your eye.

In common experience the closest most of us have come to this is probably "subliminal perception." A sensory stimulus is called "subliminal" when it is too quick or too weak for us to register it as a sensory event; and "subliminal perception" is what happens when, even so, we find that we have gone at least partway to putting a perceptual interpretation on the stimulus.

For example, we may be walking down the street and overhear a snatch of conversation or glimpse something out of the corner of our eye without—so far as we know—being aware of it at all, only to find that we now have an idea going around in our head that has come apparently from nowhere. James Alcock gives a nice instance from his own experience: "I was standing in a cinema waiting to buy some popcorn, and was idly recalling a conversation I had once had with the brother of a colleague. . . . A few moments later I turned around, and there about 30 feet away was the man himself. I recall the momentary sense of shock I felt."[9] Alcock notes that, if he had not reanalyzed his experience, he might have been tempted to attribute the coincidence to extrasensory perception. And indeed such experiences can easily be considered paranormal.

Subliminal perception was for a long time not taken seriously by psychologists, but experimental evidence has accumulated that it is a genuine phenomenon. In the visual sphere the best evidence has come from studies of "backward masking."[80] If a pattern is flashed on a screen for about a tenth of a second, a person will see it and be able to report some of its details; but if the same pattern is followed immediately by another longer-lasting pattern he will (when conditions are appropriately adjusted) no longer see the first pattern at all—as though it had never occurred. However, the first pattern may still influence his perception of the second. For example, in an experiment by M. Eagle,[81] the second pattern was a picture of a nondescript young man, while the first was a picture of the same man either wielding a knife or carrying a birthday cake. People were asked what they thought about the character of the man they saw in the second picture. Even when they were completely unaware that the first picture had occurred, they judged the second picture according to the character portrayed in the first.

Results such as these imply that—under admittedly contrived conditions—high-level perceptual processing can indeed take place despite the fact that the subject is unaware of receiving the stimulus and knows nothing about it at the level of sensation. But the phenomenon falls of course far short of a complete breakdown in sensation with relatively undisturbed perception—of the kind that might be the chronic condition of someone with the reverse of an agnosia, whose sensory channel was disabled completely by damage to the brain.

It may be fruitful to try again to imagine what it would be like. How *would* it strike you if—going about the everyday world—you were to find yourself able to answer questions about "what's happening out there" without your being able to answer questions about "what's happening to me"? The first answer presumably is that it would strike you just like that: in other words that you would find yourself able to make accurate judgments about the external world based on stimulation at your body surface without being aware that any such stimulation was occurring. But, by contrast to an agnosic patient, you would surely think something very peculiar was going on. Indeed you might well claim that whatever perceptual judgments you were making had "nothing to do with you"—and for that reason you might be loath to make them.

In the early 1970s a clinical syndrome was uncovered by Lawrence Weiskrantz that appears to exemplify just such a condition.[82] "Blindsight," as it has come to be called, occurs in certain human beings who have suffered extensive damage to the primary visual cortex that lies at the back of the brain. They are "blind" in a large part of the visual field: "blind" in the sense that they do not acknowledge that this part of the visual field exists for them at all. They say they have no sensation of light or dark or color in the blind field, just as if the corresponding part of their retinas had disappeared and the light stimulation was simply not affecting them. And yet certain perceptual faculties are still intact. If the patient can be persuaded to ignore the fact that at the level of sensation there is apparently nothing happening to him and to make *guesses* about the outside world he does surprisingly well (although by no means perfectly). If asked to reach for an object, he will reach in the right direction. Moreover if he is tested with different-shaped objects, his hand will take up the correct shape in anticipation of grasping it (try doing that yourself, and note how your fingers mold themselves to the object before they get there). If he is asked to report verbally what shape an object is he will usually fail; but if the choice is restricted to, say, an O as against an X, and he is asked to guess which one it is, he will learn in just a few trials.

I said that blindsight "appears" to exemplify the condition of perception without sensation because I do not want to overstate the case. When blindsight was first discovered, it was considered so astonishing that several commentators (including me) were tempted to make exaggerated claims about it. So let me now take a breath, and say what I really think about it in the next chapter.

· 12 ·

MORE ABOUT
BLINDSIGHT

I*have a particular* interest in blindsight, which I should relate.
Before the phenomenon was discovered in human beings, I had come across something very like it in a rhesus monkey.[83] The monkey, called Helen, was the subject of a study started by Weiskrantz in Cambridge in the 1960s. As part of his research into cortical blindness in human beings, Weiskrantz performed a surgical operation which removed almost the entire visual cortex of the monkey's brain. As a result the capacity for normal vision was totally destroyed (except possibly for a tiny tag in the far top right corner of the field of her right eye). At first this monkey simply gave up looking at things, as if she herself had no reason to believe that she could see.

I was a student in Weiskrantz's laboratory at the time and became curious about Helen. Although her visual cortex was gone, the lower visual areas of her brain were still intact, and I thought it just possible that she might have a residual capacity for vision of which she herself was unaware. I took up her case and worked with her for seven years. I coaxed her and encouraged her. I played with her and took her for walks in the fields near the laboratory. I tried in every way to persuade her that she was not blind. And slowly she did in fact begin to use her eyes again. She improved so greatly over the next few years that eventually she could move deftly through a room full of obstacles and pick up tiny currants from the floor. She could even reach out and catch a passing fly. Her 3-D spatial vision and her ability to discriminate between objects that differed in size or brightness became almost perfect.

She did not however recover the ability to recognize shapes or colors; and in other ways too her vision remained oddly incompetent. When she was running around a room she generally seemed as confident as any normal monkey. But the least upset and she would go to pieces: an unexpected noise, or even the presence of an unfamiliar person in the room was enough to reduce her to a state of blind confusion. It was as though, even after all those years, she was still uncertain of her own capacity—and could only see provided she did not try too hard to see.

Here is how I described her in 1977: "She never regained what we—you and I—would call the sensations of sight. I am not suggesting that Helen did not eventually discover that she could after all use her eyes to obtain information about the environment. She was a clever monkey and I have little doubt that, as her training progressed, it began to dawn on her that she was indeed picking up 'visual' information from somewhere—and that her eyes had something to do with it. But I do want to suggest that, even if she did come to realise that she could use her eyes to obtain visual information, she no longer knew how that information came to her: if there was a currant before her eyes she would find that she knew its position but, lacking visual sensation, she no longer *saw* it as being there. . . . The information she obtained through her eyes was 'pure perceptual knowledge' for which she was aware of no substantiating evidence in the form of visual sensations. Helen 'just knew' that there was a currant in such-and-such a position on the floor. . . . 'Blindsight' is what I think Helen had. . . . The human patient, not surprisingly, believes that he is merely 'guessing.' What, after all, is a 'guess'? It is defined in Chambers Dictionary as a 'judgement or opinion without sufficient evidence or grounds.' "[84]

But the problem about this is that it only partly fits the facts of human blindsight. For a start, the human patients never recover their vision to anything like the same degree that the monkey did. Although they can see much better than they "should" be able to, they are still not very good. It is worth contrasting the performance of Weiskrantz's star patient, D.B., with that of a blind subject wearing the skin-vision apparatus: D.B. has never even approached the level of perceptual competence that is achieved by blind subjects after only a few hours of practice with skin-vision.

But next my characterization of blindsight as "pure perceptual knowledge" such that the subject "just knows" what is in front of him is apparently contradicted by the human patient's own descriptions. Certainly the patient says he does not have visual sensation; but—rather as in the case of subliminal perception—*he* would claim he does not have perception either. He never says anything to the effect of "Gosh, isn't it strange, I just know there's something X-shaped out there, even though I can't see it." Rather he says, "*I* don't know anything at all—but if you tell me I'm getting it right I have to take your word for it." In other words it is as though he can only discover his ability at second hand, which is hardly what we might expect of "pure perception." Maybe somebody is having pure perceptions but it isn't "me"!

What is it like to have blindsight? I suggested that subliminal perception might seem like ESP, and perhaps the experience of blindsight is not dissimilar. If you have ever been the subject of a telepathy experiment using the Zener cards (which show a circle, a cross, a star, etc.) where the task is to guess which card is being telepathically transmitted from someone in another room, you will know what a peculiar situation it is. You shut your eyes and let your mind go blank, and you find maybe that the idea of a particular pattern—not exactly an image—enters your mind and you have an urge, for example, to say "cross." But, if you are a rationalist like me, you feel a bit foolish about actually claiming that you are receiving a picture of the cross—because it is quite unclear in what way the information is getting through to you (and the fact is that it is not getting through).

In the blindsight case however the information *is* getting through: and if the subject feels the urge to say "cross" it is because his eyes are really telling him there is a cross. (Actually the subject seldom feels an urge to *say* "cross" when he sees a cross; what happens, on my reading of the evidence, is that he feels an urge to *grasp* it in the appropriate way.) Even so, typically he does not believe in his own capacity and he too feels a bit foolish. Certain patients have refused to cooperate in blindsight tests for just that reason.

I claimed just now that this is not what we would expect of pure perception. But then what would we expect of pure perception?

What would someone say about it if he had it? Perhaps the fact is that pure perception, if it occurred, would never be acknowledged for what it is: the subject would always doubt what was going on, and would never be inclined to say "I just know there's something out there" because in the absence of sensation he—"I"—would not feel he had any direct personal involvement in the business of knowing.

One can get an imaginative handle on it in the following way. Try looking around the room, and then close your eyes. Naturally visual sensation will cease, since there is no longer any light arriving at the eye. But for a while at least the visually acquired knowledge of the room will persist. In fact if soon after having closed your eyes you reach for an object, not only will you reach in the right direction but your hand (without your thinking about it) will take up the right shape. It is not a case of your "just knowing" where and what shape the object is, since it is obvious to you *how* you know. You will not find your ability at all surprising.

But now imagine what it would be like if you were to keep your eyes permanently closed, and were to find that you still had knowledge of the position and shapes of objects (with this knowledge being continually updated) *as if you had closed your eyes only a moment before.* This would be a genuine case of "pure perceptual knowledge, unsubstantiated by sensation"—of "just knowing." You would now, perhaps, be in much the same situation as the monkey Helen or the blindsight patient. And it would probably be very surprising indeed.

Why is the former case not surprising and the latter very surprising? The answer seems obvious, but is revealing. In the former case you would feel confident in your perceptual judgments because you would recognize your own immediate involvement in the process of seeing; but in the latter you would have no basis for feeling that you yourself were thus involved.

Thus maybe blindsight is after all a case of pure perceptual knowledge, despite the subject's protestations that he—"I"—is not seeing in any way at all. For what seems to be strikingly lacking in the blindsight case (or subliminal perception, or for that matter ESP) is precisely this ego involvement that sensation usually provides. Possibly that is why monkeys show greater recovery than people do, since monkeys most likely have a less elaborated self-concept and hence may not be so thrown by the lack of self-involvement: foolishness is probably not an emotion that monkeys feel.

Anthony Marcel, coming at this problem from a different angle, has emphasized just the same role for sensation in *justifying* voluntary action. "People will not themselves initiate voluntary actions which involve some segment of the environment unless they are phenomenally aware of that segment of the environment [i.e. unless they have sensations]. . . . To the extent that someone is paying attention to their behaviour, they do not normally allow themselves to perform actions *without reason*."[85]

Marcel stresses, in particular, that a person with blindsight lacks such reasons—and is loath to act "unreasonably." "Consider the following situation, which has to be treated as a thought experiment since we have not carried it out in any rigorous form. If a person with cortical blindness and blindsight in one hemifield [half their field of vision] is very thirsty and a glass of water is placed such that it falls within that person's sighted field, there is little doubt that they will either reach for it and drink it or will ask if they can take it. Now suppose the glass of water is placed so that it falls within the blind field. Remember that from our own work we know that the object is apparently sufficiently well described visually to be identified and to permit adequate grasping. What will the person do? Will they do the same as when the stimulus is in their sighted field? Or will they reach out but not know why (until they contact the glass manually)? Or will they do nothing? The contention here is that they will do nothing—partly on the basis of anecdotes communicated by such people, partly on the basis of observation."

The point is not that someone in this situation *cannot* act but rather that he *does not* act. For all his life (at least until the injury) the patient has been accustomed to having his actions in relation to perceived objects "sanctioned" by the occurrence of sensation—and it seems that old habits die hard.

People can of course sometimes learn that what previously seemed to them unreasonable is reasonable after all. We have probably all undergone just such a reeducation in relation to those doors at airports that open as if by magic when we push a cart at them—without our exercising "reasonable force." Equally it might be possible for the blindsight patient to learn to trust perceptual knowledge *simpliciter*—without his having "reasonable sensory evidence." But the danger in the airport case is that one day we may push a cart at a door that does not open, and equally there would be real dangers—of a sort to be discussed in the next chapter—of acting without the say-so of sensation.

This is uncertain water, with several crosscurrents that need navigating. Nevertheless the argument is homing in on a new role for sensation in the mental economy of human beings. Sensation lends a here-ness and a now-ness and a me-ness to the experience of the world, of which pure perception in the absence of sensation is bereft.

· 13 ·

A FIRE IN THE HAND;
A DAGGER OF
THE MIND

When *Aristotle* was told that someone had been abusing him behind his back, his traditional reply was: "Let him even beat me, provided I am not there." He might have added: "or provided I 'just know' about it but don't feel it."

I must now bring affective responses back into the picture, and extend the discussion to sensory modalities besides the visual one.

Suppose there is a red-hot coal in the grate, and I reach out toward it. As my fingers approach the coal I feel a sensation of myself being scorched, and I perceive the coal out there as hot. When I take my hand away, the sensation (soon) dissipates and my fingers stop hurting—although I still know the coal is hot. Suppose, indeed, I simply look at it. I sense the light arriving at my eyes as red, and I perceive the coal out there as being red-hot. When I look away or close my eyes the visual sensation disappears and any pleasurable response to this fiery red sensation ceases—although I still know the coal is colored red.

The two cases, touch and vision, are parallel. The scorching of my fingers and the receiving of red light at my eyes are facts about me; whereas the hotness and the redness of the coal are facts about the coal. But the case of touch shows the more clearly how pleasure or pain is tied in with the presence of sensation. Although perceptual knowledge may sometimes have affective connotations, aroused through secondary associations with sensation, such knowledge by itself is *affectively neutral*.

Put like this, the point is obvious, and the explanation for it obvious too: namely that "just knowing" can have no immediate bearing on one's bodily well-being. It becomes, however, if not less obvious then more interesting when it is realized that what holds for knowledge of what is happening *somewhere else than at the body surface* holds equally for knowledge of what has happened *sometime other than the present moment.* Indeed there is no more reason why someone should feel pain on remembering being scorched by a hot coal an hour ago, than that he should feel it on knowing that there is now a hot coal a yard away.

John Locke recognized this. "Pleasure or pain," he wrote in the *Essay,* "which accompanies actual sensation, accompanies not the returning of those ideas without the external objects. . . . Thus the pain of heat or cold, when the idea of it is revived in our minds, gives us no disturbance; which, when felt was very troublesome."[86]

Poets, too, have drawn attention to the affective poverty of revived images. In Shakespeare's *Richard II,* when Bolingbroke is exiled from England, his friends attempt to comfort him by suggesting that he can always find solace in remembering or thinking about happier days. Bolingbroke replies:

> *O! who can hold a fire in his hand*
> *By thinking on the frosty Caucasus?*
> *Or cloy the hungry edge of appetite,*
> *By bare imagination of a feast?*
> *Or wallow naked in December snow*
> *By thinking on fantastic summer's heat?*[87]

O no, he says, a memory or a thought provides no comfort at all when the facts of present stimulation are so totally opposed.

Shakespeare's phrase "bare imagination" sums it up, and brings me to the more general point I want to make: namely, that not only pure perceptual knowledge but all other "unsensed ideas" (memories, thoughts, images, etc.) are *bare*—bare precisely because they lack the rich vestments of sensation. This is not to say that such unsensed ideas lack content, or even that they are entirely nonsensory in character. But it is to say that they are severely deficient in the qualitative density that sensation typically provides.

Consider an example that has become a favorite of philosophers: the "purple cow" ("I never saw a Purple Cow / I never hope to see

one; / But I can tell you, anyhow, / I'd rather see than be one"[88]).
Try imagining a purple cow, in as much detail as possible. You will
presumably have a fairly clear idea of which way the image of the
cow is facing, whether it has horns, and maybe even whether there
is a bell around its neck or a milk bucket beneath its udders.
Moreover you will have no doubt that this is a visual image (the
image of something seen) rather than a tactile or an olfactory image.
But nonetheless the purpleness of the imagined purple cow will
almost certainly be meaner, more diaphanous, more fleeting than
any real-life purple that you ever saw: to imagine a purple cow is
just not the same thing as to have a purple sensation (or at least a
purple sensation worth the name).

Or, to change modalities, consider hearing a voiced thought in
your head. Suppose you think to yourself in words: "The rain in
Spain falls mainly in the plain." You could probably say in whose
voice the imagined words are spoken (it is most likely your own
voice, but it might for example be Audrey Hepburn's voice as you
remember it from *My Fair Lady*), you could describe the manner of
pronunciation (Queen's English or a cockney accent), and be able to
confirm that the words still rhyme. You will have no doubt that the
image is an auditory image (the image of something heard). But
again, as with the purple cow, the imagined sounds will not have the
density of real sounds arriving at the ears.

Compare imagining the sounds "swish, swish" with actually hearing
them (the reason for choosing this example will become apparent in
a moment). The two experiences are presumably not equivalent. Yet
it might be possible, at least in principle, to contrive circumstances
in which the two experiences *would* be equivalent.

Here is a genuine case history.[89] In 1928 a patient turned up in a
Boston hospital who had been born with a large cluster of abnormal
blood vessels at the back of his brain in the region of the visual
cortex. To the doctors' amazement he told them that whenever he
opened his eyes he heard a swish, swish sound, like the noise of wind
blowing in his ears. But this was not imaginary hearing, it was the
hearing of real-life sounds. And when the doctors put a stethoscope
to the patient's scalp, they too could hear the swish, swish. The
sound would begin for example when the man started reading a
newspaper, and stop when he closed his eyes.

The explanation, though unusual, was straightforward. Whenever the visual cortex becomes active upon receiving stimulation from the eyes, there is—in everybody—an increase in blood flow to this part of the cortex (to help it, as it were, to carry out the extra work). In this particular man, however, the increased blood flow went to the abnormal vessels; and as it rushed through these vessels it made an audible sound. So the man was, in effect, "hearing himself see."

Now, in the light of this, one might invent another case: that of a patient who had been born with similar abnormal blood vessels in the region of his auditory cortex instead of his visual cortex. Whenever this man started listening to an external sound he would presumably hear the swish, swish of the blood rushing to the vessels in his auditory cortex (that is, he would hear the swish, swish as well as the original external sound). He would, in effect, be "hearing himself hear."

So now comes the crucial thought experiment. It is known (as will be discussed in the next chapter) that the visual or auditory cortex becomes active not only when there is external stimulation via the eyes or ears but also when the subject merely *imagines* a sight or a sound. Thus the Boston patient would presumably have heard swish, swish even if he had just imagined looking at a newspaper (although this was never tested), and likewise our new patient would hear it if he were just to imagine hearing an external sound. But then suppose the sound he were to imagine hearing were to be "swish, swish": *he would find himself hearing, as a real-life sound, the very sound that he imagined.* Hence (perhaps for the first time in human history) we would have a man whose self-generated image of a sound was accompanied by the full-blown sensation of that sound arriving at his ears.

This case is so silly that I trust it makes its point: namely that this patient's situation is altogether unlike that of normal people.

The "feel"—or, rather, the lack of it—of imagery gives it, perhaps, a somewhat puzzling status. But there is not really so much to be puzzled about. In fact, if one thinks biologically, it would surely be much more puzzling if people did generally experience full-blown sensations in association with mere images, memories, or thoughts.

Sensations, as we have seen in previous chapters, have a well-defined biological role as representations of "what is currently happening to me as an embodied being." Sensations prepare the subject to take immediate action to extend or escape or ameliorate his present situation vis-à-vis the stimulation arriving at his body surface. And it would clearly be an error biologically for a person to represent an image of what might be happening to him—if it were sometime else or if he were somewhere else—as a *present* sensation. If a human being could—and therefore very likely would—actually hold a fire in his hand by thinking on the frosty Caucasus or satisfy his appetite by bare imagination of a feast, the chances are he would end up starving and covered in blisters. Natural selection would presumably eliminate him (and any other such fantasist) within a generation.

There are therefore excellent evolutionary reasons why imagination should be relatively bare. It is essential that, when and if a person conjures up images of nonpresent stimulation, he should mark these images off as being "not-for-real." And the very absence of sensation can immediately achieve this: serving, as it were, to put scare quotes around such images—to indicate "this is not what it might seem."

I mentioned Shakespeare's Bolingbroke; but what about Macbeth. When Macbeth, in the play, has a vision of a dagger, he reaches out for it and finds himself grasping empty air:

> Art thou not, fatal vision, sensible
> To feeling as to sight? or art thou but
> A dagger of the mind, a false creation,
> Proceeding from the heat-oppressed brain?[90]

For Macbeth the unreality of the image is revealed when he fails to get the expected feedback from his hand. But Shakespeare's lines might better still describe the usual case where ordinary images are quickly revealed as creations of the brain by the test of whether they are "sensible" at all.

In general, if a person is ever in doubt about whether what he is seeing relates to what is physically present, he can always check up by asking himself: "What does it feel like at the level of visual sensation?" If the answer to this question is "It does not feel right"—in other words he is not having the sensations he would expect to have—he can be pretty sure his mind is wandering.

. . .

The exceptions prove the rule. The last chapter focused on the case of blindsight, where the patient does *not* trust the *valid* information his eyes are giving him because it does not feel right. But more familiar are those cases where ordinary people *do* trust *invalid* information for the opposite reason. In dreams, for example, the dream images are for many people "sensible": that is they are accompanied by the full richness of sensation—and colors, sounds, smells, touches, sexual stimuli are experienced as if the stimuli were directly impinging on the dreamer.

"Dreams," Samuel Coleridge wrote, "are no shadows with me; but the real substantial realities of life."[91] And when this is so, the affective responses are present too. Hence the dreamer may cry out with fear or have an orgasm or weep wet tears, although these reactions are (biologically) quite inappropriate. What is more, the dreamer would initiate voluntary actions if only he could do so—and it is only because in the dreaming state his voluntary muscles are effectively paralyzed that he stays where he is.

In waking hallucinations resulting from pathology or drugs the same is often true, so that the hallucinator may fight against an imaginary torturer, or be disgusted by an imaginary smell, or hide his eyes from the glare of the glory of the Lord—and here the consequences are liable to be more serious because he is free to move.

It is fortunate—which is to say evolutionarily well managed—that most waking imagery does not have this up-front sensory quality. For it means we can play, as it were, with memories, images, and thoughts without surrendering our hold upon the reality of the present moment.

The very word "present" comes from the Latin *prae-sens*. *Prae* means "in front of" and *sens* is the present participle of *sum* ("I am"). But *sens* is also the root of the past participle of *sentio* ("I feel"). Thus *sens* hovers ambiguously between "being" and "feeling," and *prae-sens* carries the implication of "in front of a feeling being." Correspondingly the subjective present is comprised of what a person feels happening to him; and when he ceases having sensations—as when he enters dreamless sleep or dies—his present ends.

· · ·

I wrote that there was nothing very puzzling about all of this. But there is nevertheless still something rather puzzling about the way that we experience images. Granted that images do not involve full-scale sensations, nonetheless they do seem to have some sort of sensory component to them: they do seem to involve rather more than "just knowing."

Think back to the example of looking around the room, closing your eyes, and then reaching for an object. The fact that in this situation you reach correctly demonstrates that you do indeed know where and what shape the object is: but you probably do not—and certainly need not—have a concurrent visual image. By the same token, when I suggested that blindsight may be a paradigm case of "just knowing" I certainly did not mean to suggest that what it is like to have blindsight is the same as what it would be like to have a continuous stream of visual images. Were it like *that* to have blindsight, presumably the patient would tell us so—but he tells us nothing of the kind.

But if just knowing that something is happening out there is less than there is to having an image of it, while sensing something happening at my body surface is more, where in the scheme of things *do* images fit in?

Since there is as yet no generally accepted theory of imagery, the way is open for a hypothesis that otherwise I might be shy of putting forward. I shall describe this hypothesis in some detail in the next chapter. For I need a theory of imagery. Otherwise, when I get to talking about consciousness, I shall find myself as embarrassed as every other commentator seems to have been by the problem of where to put those mental representations that, between sensation and perception, are neither fish nor fowl nor good red herring.

14

HE THOUGHT HE SAW
AN ELEPHANT

To *explain this* hypothesis about the nature of imagery I must return to some preliminary considerations about the difference in the epistemic status of sensation and perception: that is, their status as bearers of knowledge of the facts.

Here, again, is the basic diagram of parallel sensory and perceptual channels:

$$object \rightarrow light\ at\ retina \underset{\text{perception of what is happening out there}}{\overset{\text{sensation of what is happening to me}}{\bigg\langle}}$$

We can take it, presumably, that whenever someone is being stimulated at his body surface there can be said to be a truth of the matter about "what is happening to me" and about "what is happening out there." When, for example, I look at a red coal *there is* a particular pattern of stimulation on my retina and *there is* a particular physical object out there.

However, a person's means of access to the two kinds of *there is*, via sensation and perception, are clearly not on a par. The process of sensory representation need not involve much more than making an internal *copy* of the physical stimulus as it is occurring at the body surface; but the process of perceptual representation has to involve something more like making up a *story* about what this stimulus signifies to be occurring in the outside world. Thus, while sensation

Figure 2

provides relatively direct and certain knowledge about "what is happening to me," perception can provide only relatively indirect and conditional knowledge about "what is happening out there."

An elementary illustration shows up this difference between copying and storytelling. Figure 2 is the famous wife/mother-in-law devised by Edward Boring. If you attend to perception, and concentrate on what the picture signifies as lying out there, you will probably find that your perceptual channel comes up with one of two alternative stories: either you will perceive a young woman (in profile to the left) or an old lady (with her chin buried in her fur collar)—and as you continue to look, the story may switch from one to the other. But if you attend instead to sensation, and concentrate on what you sense to be happening at your eyes, you will find that your sensory channel provides you with an unambiguous representation of a particular pattern of black and white light.

In general, perception involves a lot more complex information processing than does sensation. We might therefore well expect that the ways the brain goes about the two tasks would basically be different. And, although too little is known to be sure, there are good grounds for supposing that the sensory channel makes use of "analog" processing and ends up with a pictorial representation (something like a picture in the brain), while the perceptual channel makes use of "digital" processing and ends up with a propositional representation (more like a description in words).

However this may be, perception undoubtedly requires more *ad hoc* assumptions and more risky calculations than does sensation—the cup of stimulation and the lip of representation are more widely separated. And for that reason it is inevitable that perception is much more likely than sensation to slip up.

Fortunately (for reasons that will be considered shortly) the slips of perception, under normal circumstances, are generally not serious. But for evidence of the potential dangers, we have only to recall what may occur when the perceptual channel is for one reason or another not functioning well. For example, people suffering from visual agnosias of the kind described earlier make perceptual judgments that are not merely inaccurate but nowhere near correct. An agnosic patient may perceive a pair of scissors to be a comb—and when asked how to use the object in question will mime drawing it through her hair. Oliver Sacks has described a patient who, notoriously, mistook his wife for a hat.[92]

When the sensory channel malfunctions, however (provided it does not break down altogether, as it does in blindsight), the errors tend to involve distortions of the sensory field rather than outright mistakes. In cases of what is called visual "metamorphopsia" a patient may have the impression that parts of the visual image are swelling or shrinking, or that colors are washed out—but the overall topography of the field remains more or less intact.[93]

These different patterns of error are just what we might expect if the two channels do indeed employ very different styles of information processing—digital as against analog, propositional as against pictorial. Consider, as an analogy, the game of Chinese Whispers. If people sit in a circle and pass on a message in words (i.e. propositionally), a single small error can result in major changes in meaning—"the life of man is nasty, brutish and short" becoming, say, "the wife of man is nasty, brutish and short." But if, instead, they were to pass on a copy of a drawing, a single small error would be likely to prove relatively insignificant—a map of Britain, for example, would probably remain recognizably a map of Britain. Perception, in terms of the risk it carries, is much more like the first game; sensation like the second.

Now, there can be no doubt that perceptual error, if it generally went uncorrected, would prove a biological calamity. The man who regularly mistakes his wife for a hat (or worse still, his hat for a wife) is headed for extinction.

Something therefore must be done about perception. In the course of evolution there must have been strong selection pressures in favor of developing some sort of error-detecting mechanism: some way of checking the results of a perceptual calculation before acting on it. And the fact that today most people in normal circumstances do not on the whole make wild perceptual errors strongly suggests that a natural solution to this problem has indeed been found.

It is important to examine what this solution may be—not only because of its intrinsic interest but because it holds the key to further developments.

Suppose you were asked: "What is the square root of 143641?" If you know how to calculate square roots, you would eventually arrive at the answer 379. But suppose you were worried that you might have made a mistake in your calculations. Then the obvious way of checking on them would be to go into reverse and to ask yourself: "What is the square of 379?" Provided you ended up with the number that you started with, you could be pretty sure your answer was correct. Indeed if all you wanted was a rough and ready check, you might simply observe that since the last digit of 379 is a 9, and the square of 9 is 81, 379 can be the square root only of a number ending with a 1. By just squaring the last digit of your answer, you could in fact quickly detect on average 80 percent of all random errors.

This strategy of "echoing back to the source" is a strategy of error detection well known to information technologists, who under a variety of circumstances may need to check that an operation has been performed correctly, or that a message has been decoded in the right way, or just that a signal has got through on a noisy channel. The trick is to *undo* the operation, *recode* the message, or send the signal *back to its source*—and in each case to compare the reconstructed data to the original. The procedure can be called the "Grand Old Duke of York strategy" (after the Duke who "had ten thousand men,/he marched them up to the top of the hill/and he marched them down again").

Then why should not a version of the Grand Old Duke of York strategy be employed in the case of perceptual processing? The perceiver might start by asking himself, in effect, "What does this stimulus on my retina correspond to in the outside world?" After a series of complicated calculations he would reach the answer, perhaps "a hat." But then, just to make sure that he has not slipped up, he would attempt to reconstruct the original retinal stimulus from the perceptual representation. If the reconstructed stimulus turned out to match the original stimulus, well and good; but if it did not match—because, say, the original stimulus was produced by light coming from a wife and not a hat—something must have gone wrong.

This strategy would not catch *all* perceptual errors, because sometimes an erroneous perceptual conclusion might accurately reflect the original data. But it could at least be counted on to catch the worst of them. And provided only partial reassurance is required, the same shortcut might be available as with the numerical example. Thus the perceiver might feel safe enough if he were simply to reconstruct a "rough and ready" version of the stimulus and match it to the original: not, for example, the hat in all its richness, but a cartoon or outline version, sufficient at least to provide a mismatch with a wife.

It is true that the reconstruction even of a cartoon version of the original stimulus would not be possible unless a great deal of otherwise redundant contextual information about the perceived object was preserved: e.g. about where it lay in relation to the direction of gaze, how far away it was, and so on—none of which is directly relevant to the object's being a hat. But there is every reason to suppose that such contextual information is in fact available at the perceptual level.

When we perceive a hat, we perceive both what it is and where it is: and the fact that we are able to reach for it accurately, shaping our hand exactly to its contours, shows that we must have preserved all the relevant information about how it lies in relation to our body. Indeed when, in order to grasp a perceived object, we send command signals back to our fingers we must be doing very much the same job of back calculation as that which would be required to reconstruct the retinal stimulus—in each case regenerating an analog description from a digital one.

If this is, in principle, the strategy that is employed, how might it be put into practice in the brain? In particular, *where* might one expect the comparison of the reconstructed stimulus with the original stimulus to take place?

One possible answer would be: at the sense organ itself. Thus, in the case of vision, the information starting at the eye might be marched up to the "perceptual center" in the brain and then be marched back to the eye. But this is implausible for several reasons, not the least of them being that by the time the reconstructed stimulus got back to the eye the original stimulus would very likely no longer be present—because, for example, the eye had moved.

There is however an obvious and better alternative that could serve as the locus of comparison: namely the place in the brain where, as it happens, an extant copy of the original stimulus has already been prepared—in other words wherever it is that the *sensory representations* are held. So the "perceptual center" might well send its reconstruction of the stimulus straight over to the "sensory center," where a comparison could take place with what is there already.

The scheme, then, would be something like this:

$$
\begin{array}{l}
\text{sensation of what is} \\
\text{happening to me} \\
\\
\text{object} \rightarrow \text{light at retina} \prec \qquad \uparrow \\
\text{perception of what is} \\
\text{happening out there}
\end{array}
$$

If there was a good-enough match, the perceptual representation would be accepted; otherwise it would have to be immediately revised.

It will become clearer in a moment how all this could bear on the question of imagery. But first, by way of a diversion from too much theory, I can cite some tantalizing evidence that something like this is occurring in the human visual system: evidence of, apparently, a "top-down influence" of perception on sensation.

Figure 3 shows the "tabletop illusion" (an illusion which, surprisingly, went unrecognized till twenty years ago[94]). The table appears to be drawn in reverse perspective, with the far side longer than the near side. But if you put a ruler to it, you will find the tabletop is *drawn* as a perfect parallelogram, with opposite sides equal.

Figure 3

Notice that this is an illusion at the level of sensation, as well as of perception. It is not just that the far side of the perceived three-dimensional table looks to be longer than the near side, it is that the image of the upper line as it is being sensed as a retinal stimulus appears to be longer than that of the lower line.

So, what explanation might apply? Admittedly nothing in what I have proposed so far would entail that the signal sent over from the perceptual center to the sensory center could actually *modify* the sensory representation of the visual stimulus. But it is easy to believe that if two representations of the same stimulus are arriving in the same place, there might be some degree of interaction between them.

Perhaps what is happening is this. The perceptual center, applying the laws of linear perspective, is making a correct three-dimensional interpretation of the drawing: as a table in which the far edge is both further away and longer than the near edge. To check on this interpretation it is then trying to reconstruct the visual stimulus by, among other things, undoing the perspective. However, it does not undo the perspective quite enough, with the result that the reconstructed version of the retinal stimulus that is sent over to the sensory center has the upper line slightly too long. The match however is so nearly correct that, instead of the perceptual representation being rejected, the sensory representation is itself getting modified to bring it into line.

Similar illusions, with similar interpretations, were studied in the 1930s and 1940s by experimental psychologists interested in what is called "perceptual constancy." They discovered that there is a general tendency in vision for the sensory representation of the stimulus to be pulled, as it were, toward an "ideal" view of the external object—as if the object were being seen full-frontally.

Figure 4, for example, is a diagram from a classic paper by Robert Thouless.[95] It shows how an inclined circular disc looks to an observer when he attends to his sensory experience. The subject was required to look at the disc laid flat on a table and to match its "appearance" to one of a series of ellipses held up vertically in front of him. Thouless comments that, to make sure the subject understood what was required of him, "I have generally given him preliminary practice . . . pointing out that I want to know neither what the shape of the object really is nor how he thinks it ought to look but simply the shape it does look to him. Even the most ignorant subject understands these instructions perfectly." The results showed that subjects consistently judged the appearance of the elliptical stimulus at the retina to be more circular than it should be.

Thouless came up with a general name for this effect: "phenomenal regression to the real object." "Phenomenal," as he used the term, meant in the realm of sensation, while "real object" meant in the realm of perception. He stated "the law of phenomenal regression" to be that "when a stimulus which by itself would give rise to a certain phenomenal [i.e. sensory] character is presented together with perceptual cues which indicate a "real" character of the object, the resulting phenomenal character is neither that indicated by the

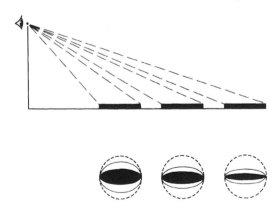

Figure 4
Real shape shown by circle, retinal shape by solid ellipse, "phenomenal shape" by outline ellipse

stimulus alone nor that indicated by the perceptual cues, but is a compromise between them."

Without a scheme of the kind proposed above (though not necessarily this precise one), the implied downward influence from perception to sensation would be baffling.

Everything discussed so far in this chapter has concerned perception and sensation in the presence of stimulation from external objects. But it is now an easy step to relate it to self-generated images.

It may help focus the argument if I describe (as best I can) my own experience when, for example, I try to imagine a purple cow. To make it more taxing, though more typical, let me assume my eyes are open and I am looking out the window at the cloudy sky—so that a competing stimulus is arriving at my retina.

The experience is not easy to put into words (and it may not be everyone's experience), but roughly it seems to be this. I "see" a fleeting image of the cow, coming and going on top of the patterned visual field produced by the cloudy sky. And this experience of "seeing" is made up of several elements. At the level of perception, what I *perceive*—so long as I can hold on to it—is in fact dominantly a representation of a cow (I could describe the color of its skin, the shape of its ears, the lie of its tail); and, while this is going on, I hardly perceive the clouds as an external fact. But at the level of sensation, it is much more complicated. The external field is still there, and what I *sense*—even while I am holding on to the image—is dominantly the retinal stimulus produced by the light coming from the cloudy sky (I am aware of the blobs of color and so on). But in addition to this I have what I can only describe as a will-o'-the-wispish impression of a patchy, cow-shaped, purplish-colored projected image—a version of the retinal stimulus I would be receiving from a purple cow *if* it were now before my eyes.

To explain this experience, in the light of the scheme I have been developing, all that has to be added are four plausible proposals.

(i) Images arise in (or at least come via) the perceptual centers of the brain.

(ii) When the perceptual center is engaged in generating images, it is temporarily disengaged from normal perception.

(iii) When the perceptual center generates an image, the attempt to check for perceptual error still goes on, even though in reality there is nothing to check it against. Thus there is an attempt to reconstruct the original stimulus that would have been produced by that object *if* it had been stimulating the sense organs (under "standard" or "ideal" conditions); and this reconstruction is sent over to the sensory center.

(iv) The reconstructed stimulus does not match the stimulus actually arriving at the retina. Hence the imagined representation is *rejected*. And for that reason it is extremely difficult to hold on to the image.

Thus, in the case of my particular example, the diagram below tells the story.

sensation of
what is happening to me

cloudy sky → light at retina ↗ ↑

idea of purple cow
out there

Light from the cloudy sky stimulates the retina, and gives rise to sensations in the usual way; however, it does not give rise to perceptions because this channel is temporarily closed down. Instead, on the perceptual side, the perceptual center generates its own representation of a cow, as an idea of what *might* be happening out there. The perceptual center then runs a check on this self-generated representation by attempting to reconstruct the stimulus that a real cow in front of the eyes would have evoked, and this is sent over to the sensory center. But the reconstructed stimulus does not match. So the image is rejected and the imagined cow continually fades away, and has to be renewed.

In speaking of my own experience of the image, I said that the sensation corresponding to the genuine retinal stimulus is "dominant" over the sensation corresponding to the imaginary stimulus. As an analogy (which may be more than an analogy) for this kind of "dominance," consider the phenomenon of binocular rivalry, such as occurs in ordinary vision when two noncompatible images arrive at the two eyes. For example, while looking at this page, place the forefinger of your right hand in front of your right eye, close to your face. You will probably find yourself seeing the page through a transparent "ghostly" finger. Because you are focused on the page, the stimulus to your left eye is dominant and the corresponding sensation is completely filled in, but nonetheless the stimulus to the right eye is still being registered after a fashion.

Binocular rivalry occurs when there is competition between two sensory representations corresponding to two different stimuli actually arriving at the two eyes. But it would seem entirely plausible that there should be similar rivalry in the case of imagery, when there is competition between a sensory representation of a genuine stimulus and a reconstructed representation of an imaginary stimulus.

Imagined cows are certainly not much like ghostly fingers. But they are surely a bit like ghostly fingers. (And, of course, imagined ghosts—for those who see them—are quite like ghostly fingers.)

In the situation just described, with your forefinger in front of your right eye, shut your left eye. Now, of course, the stimulus to your right eye, having the field to itself, becomes dominant in its own right and your finger suddenly appears "solid." If the analogy holds, we might expect that the sensory vividness of imagery would also be considerably enhanced if and when the image were to have the field to itself—for example, if there were to be no external stimulus arriving at the eye.

Most people would agree that it is easier to create a strong visual image if they look at a blank wall, or better still if they close their eyes or go into the dark. John Donne, in recognition of this, wrote: "Churches are best for prayer, that have least light: / To see God only, I go out of sight."[96] However, even deliberately going "out of sight" may prove less than enough to bring about the complete absence of visual sensation at the eyes. For what may be experienced instead is the positive presence of blackness: the sensation that "there is *no* light arriving at my eyes." And this sensation of blackness generally overrides in sensory richness any self-generated image.

For there to be no competition at all from the sensory representation of the stimulus at the retina, there would presumably have to be *no such* sensory representation. And the only situation where that would be likely to occur would be when the input from the eyes to the brain was actively blocked—*as happens when a person is asleep.*

What, then, about images that are generated in a sleeping brain? What about dreams? The difference between dreams and waking imagery lies, I suspect, precisely in this. When someone is asleep, no signal from the retina reaches either the perceptual or the sensory centers, and so dream imagery literally has the field to itself.

In the diagram now, the whole of the left-hand side can be left out.

sensation of
what is happening to me

. . . .   ↑

dream of what is
happening out there

When a "dream idea" arises, the perceptual center generates a representation of the appropriate external events and then attempts to run a check on its own representation by reconstructing the stimulus to which the dream events would have given rise if they were really happening. But now the reconstructed stimulus is not in competition with any other sensory representation, and hence it can dominate sensation—with the result that the dream image is experienced as exceptionally rich. Moreover, because there is nothing to indicate any kind of mismatch, there is now nothing to tell the perceptual center to revise its calculations—with the result that the dream image does not vanish as soon as it is formed.

In the case of waking imagery all images are, in effect, treated as "mistakes" and this is why they do not last for long. But in the case of dreams, even if there were to be an actual mistake in translating a dream idea into a perceptual representation, the mistake would presumably go uncorrected for some time. The consequences might be just as we experience them: not only would dream images be more vivid and less fleeting than waking ones, but they would also be more liable to bizarre digital-processing-style errors. If, for example, when dreaming of a wife, the perceptual center were mistakenly to generate a representation of a hat, the dreamer might find himself thinking of a wife while experiencing the image of a hat: and so it would remain, until perhaps some random event rejigged the perceptual calculation.

Lewis Carroll's story *Sylvie and Bruno* catches exactly this oddity of dreams.[97] Some verses from the "Gardener's Song" can fittingly round off this discussion about imagery and perceptual error:

> He thought he saw an Elephant,
>   That practised on a fife:
> He looked again, and found it was
>   A letter from his wife.
> "At length I realize," he said,
>   "The bitterness of life!"
>
> He thought he saw a Rattlesnake
>   That questioned him in Greek,
> He looked again and found it was
>   The Middle of Next Week.
> "The one thing I regret," he said,
>   "Is that it cannot speak!"
>
> He thought he saw a Banker's Clerk
>   Descending from a bus:
> He looked again, and found it was
>   A Hippopotamus:
> "If this should stay to dine," he said,
>   "There won't be much for us."

I shall not however round it off quite there: because there is some scientific evidence that I have been holding in reserve.

If imagery involves a signal sent back to the sensory center, then—if the scheme above is to be taken literally—it would imply that one and the same area of the brain must be active both when a person is sensing an external stimulus and when he is generating an internal image.

Now, in the case of vision, we know that when light falls directly on the retina of the eye, there is activation of a corresponding region of the primary visual cortex at the back of the brain. Moreover direct electrical stimulation of this area of cortex in an awake human subject causes him to have sensations of light, and when this area is damaged (as in blindsight) light at the eye no longer gives rise to sensations at all. We can conclude therefore that the primary visual cortex certainly forms part of the sensory channel. All the same, this area of cortex is only two nerve cells away in a direct line from the eye itself, and it might seem extremely implausible to suppose that it is what I have been calling the sensory center—as it were the seat of visual sensations—let alone that it could be directly implicated in the generation of visual images.

It is all the more remarkable therefore that recent physiological studies have shown that self-generated visual imagery does in fact produce activation of the visual cortex. The evidence comes from studies both of electrical activity of the brain and of cerebral blood flow while subjects undertake such varied tasks as visualizing going for a walk, imagining a cat, and answering a question such as "Is the green of pine trees darker than the green of grass?" Martha Farah has reviewed these studies (which include her own) and concludes that "across a variety of tasks, it has been found that visual imagery engages visual cortex, whereas other tasks which are highly similar save for the visual imagery, do not."[98] Moreover, as Farah notes, these findings are complemented by evidence that when the visual cortex is damaged, there is a loss not only of externally produced visual sensation but of visual imagery too.

This finding is certainly dramatic. With just a little science-fictional license, it would seem possible that, when someone imagined a cat, the image of the cat would be "back-projected" on the retina (where it could be "seen" by someone else!). This possibility is not of course the reality. But the reality is very surprising. And to explain it, a hypothesis no less surprising than the "Grand Old Duke of York" hypothesis is called for.

15

H E R E   I T   L I E S

I *have been creeping* up on the big question of consciousness.

Earlier, when I said that Aristotle's reply of "Let him even beat me provided *I* am not there" could equally have been "or provided I just know about it but don't feel it," I was already getting near: for I might have said "or provided I am not conscious at the time." And before that, when discussing blindsight, I came even nearer: for several observers have claimed that the blindsight subject, who lacks visual sensation and insists that he is not a present participator in his own perceptual processes, is "not conscious" of seeing.

In fact, the general area where consciousness is lying has been becoming more obvious by the chapter. And the goal must now be to lift it clear of the water and get it to dry land—before examining what has been caught at greater leisure. It is however a notoriously slippery quarry, and were I to have snatched at it too soon—before dealing with the problem of imagery in particular—I might still have ended empty-handed.

The time has now come to make a series of quick moves. Drawing on everything discussed so far, a case can be made for the following assertions:

1. To be conscious is essentially to have sensations: that is, to have affect-laden mental representations of something happening here and now to me.

2. The subject of consciousness, "I," is an embodied self. In the absence of bodily sensations "I" would cease. *Sentio, ergo sum*—I feel, therefore I am.

3. All sensations are implicitly located at the spatial boundary between me and not-me, and at the temporal boundary between past and future: that is, in the "present."

4. For human beings, most sensations occur in the province of one of the five senses (sight, sound, touch, smell, taste). Hence most human states of consciousness have one or other of these qualities. There are no nonsensory, amodal conscious states.

5. Mental activities other than those involving direct sensation enter consciousness only insofar as they are accompanied by "reminders" of sensation, such as happens in the case of mental imagery and dreams.

6. This is no less true of conscious thoughts, ideas, beliefs . . . Conscious thoughts are typically "heard" as images of voices in the head—and without this sensory component they would drop away.

7. If and when we claim that another living organism is conscious we are implying that it too is the subject of sensations (although not necessarily of a kind we are familiar with).

8. If we were to claim that a nonliving organism was conscious, the same would have to apply. A mechanical robot for example would not be conscious unless it were specifically designed to have sensation as well as perception (whatever that design involves).

· 16 ·

HERE WHAT LIES?
A CHAPTER
ABOUT DEFINITION

As if to remind me, on cue, of the trouble this discussion may be running into, I have just received in the mail a manifesto for a forthcoming workshop on consciousness.⁹⁹ The author of it, Aaron Sloman, opens his remarks: "The noun 'consciousness' as used by most academics (philosophers, psychologists, biologists . . .) does not refer to anything in particular. This implies for example that you cannot ask how it evolved, or which organisms do and which do not have it."

The last thing I wished to do, at this critical juncture, was to get involved in an arid discussion about definition. But since there is no chance of establishing any one of the assertions set out in the previous chapter unless we have a common understanding of what their verbal content is, and since I shall eventually want to ask precisely the questions that Sloman says cannot be asked, I must now try to show not just that consciousness *can* be defined as referring to "something in particular" but that consciousness *has already* in effect been defined as "something in particular"—if not by Sloman's academics, then by ordinary speakers of the English language.

The task may not be simple. Whatever "consciousness" means now, there is no denying that it has in the past had a range of different meanings, and some of the earlier meanings are still extant. To set the scene, therefore, it will be worth making an excursion into etymology in order to examine the word's curious history. "Words," as Aldous Huxley observed, "are the instruments of thought; they form the channel along which thought flows; they are the moulds in which thought is shaped.''[100] And the reverse is also true: thoughts are the molds in which words are shaped, they form the channel along which words flow: words come into use or change their meaning as and when people have a prior idea they are striving to express.

The word "conscious" derives from the Latin *con,* meaning "together with," and *scire,* meaning "to know." In the original Latin the verb *conscire* (from which came the adjective *conscius*) meant literally to share knowledge with other people. This implied, originally, sharing the knowledge widely. But, as time went on the usage changed, and it shifted to mean sharing knowledge with some people but not others, sharing it within a small circle—and thus being privy to a secret. Caesar and his generals, for example, were *conscius* of their battle plans.

There was then a further change in this direction. The circle of those with whom the knowledge was shared became drawn tighter and tighter—until eventually it included just a single person, the subject who was conscious. To be *conscius sibi,* conscious with one-self, had come to mean that the subject was the only one who knew something—and by implication that he was unwilling to share it with anyone else. By the first century A.D. Horace could write that a man's epitaph should be *"nil conscire sibi"*: to be "conscious with himself of nothing," and so to have no guilty secrets.

After the word "conscious" came into English in the Middle Ages, its meaning underwent another shift. People wanted to make a distinction between, on the one hand, "having private knowledge that one would not want anyone else to have access to" (for example—as already implied by Horace—knowledge of one's own secret actions), and, on the other hand, "having knowledge that by its very nature no one else could have access to" (for example, knowledge of one's innermost thoughts and feelings). The work was therefore split between two words. Guilty knowledge, that was only contingently private, became something on a person's "conscience," while self-knowledge, that was more necessarily private, continued to be something of which a person was "conscious."

So, by the seventeenth century, Shakespeare could write: "The play's the thing wherein to catch the conscience of the king" (it being on the king's conscience that he had killed Hamlet's father); while in the same century Locke could write that "a man is always conscious to himself of thinking . . . consciousness is the perception of what passes in a man's own mind."

It is true that, even in modern usage, there are occasions where the more archaic meanings are preserved (and this is especially true in other languages than English). If someone were to say, on being given an award for bravery, "I am conscious of the great honor being done me," he might well mean, "I am aware of it along with you"; if, in writing a newspaper editorial, he were to talk of "national consciousness," he might mean a shared conception of belonging to a special group; if he were to say, in the confessional, "Father, I am conscious I have sinned," he might mean that it was on his conscience. But, leaving aside such special contexts, it is clear that by far and away the more common modern English meaning of "to be conscious" is to have knowledge of one's own private feelings and thoughts. Most of the earlier uses are not only no longer current, they are not allowable.

Indeed today it would generally no longer be considered either natural or correct (although it might be understood) to say "I am conscious of" about anything other than a personal fact: I might say "I am conscious of having a toothache," but not "I am conscious of Paris being the capital of France." Nor would it be natural to say it about a personal fact other than a fact relating to myself: "I am conscious of *my* having a toothache," but not "I am conscious of *your* having a toothache." Nor about a fact relating to myself other than when the evidence for it is now before my own mind: "I am conscious of my *having* a toothache now," but not "I am conscious of my *having had* a toothache yesterday."

Thus, as the English language has evolved (and perhaps as the users of the language have become more self-concerned and introspective) the meaning of the word "conscious" has not only become narrower and narrower, it has in effect turned around. Rather like the word "window," which has changed its meaning from "a hole where the wind comes in" to "a hole where the wind does *not* come in," "conscious" has changed from "having shared knowledge" to "having intimate knowledge *not* shared with anyone except oneself."

Moreover, during the last two centuries there has been another major shift of emphasis: from using the word "conscious" *transitively*, "I am conscious of such and such a thing . . . or conscious that such and such is the case . . ." to using it *intransitively,* simply "I am conscious [stop]" or "he or she is conscious [stop]"—where conscious now denotes a special state of being. This opened the way to distinguishing "consciousness" (the state of being conscious) from "unconsciousness" (the state of being not conscious). And increasingly, over the years, the focus of discussions about consciousness has come to be on this distinction.

This history may not be acknowledged (and to modern users may not matter). Nonetheless, I take it to be beyond dispute that the word "consciousness," especially in its later intransitive sense, is now an established part of our vocabulary. And, even if ordinary people do not use the word every day, most seem confident enough about its scope and limits. Not only do they use the word in the same place in the same sentences, but they frequently concur on the truth value of such sentences. If you doubt that, try these for size on your own lips: "the patient regained consciousness when the anesthetic wore off," "you can't deny that chimpanzees are conscious," "the astronauts lost consciousness before the space shuttle hit the sea," "you can't enjoy sex if you're not conscious," "my computer has no moral rights because it isn't conscious," "although I lose consciousness when I fall asleep, I am conscious when I'm dreaming," "there couldn't be art without consciousness," "Louis XVI remained conscious for at least ten seconds after his head was cut off." Even if you would not in fact agree with all of the preceding statements, I have little doubt you understand them.

But *what* do you understand by the word "conscious" in these different statements? And is your understanding the *same* in every case? It is my purpose to show that in (almost) every case there is at least an implicit assumption that "to be conscious" is indeed essentially "to have sensations"—or more generally "to have affect-laden mental representations of something happening here and now to me."

To do this, I shall argue as follows. First, that "the having of sensations" is a naturally demarcated and psychologically significant state, with the right sort of credentials to fill the bill. Next that people grow up to recognize this state as a natural kind, and from early childhood employ it as a conceptual tool for categorizing the condition of living (and nonliving) things. Then that the name for this kind of state in the English language is—or has come to be—"consciousness." Last that when people talk about the "mystery of consciousness" or speculate for example about whether animals are conscious, it is almost always this particular meaning of consciousness they have in mind.

For the first steps I am going to enlist some innocent help.

I recently asked an eight-year-old girl, Lily, what "consciousness" means: she gravely informed me that Yes, she had heard the word, but No, she did not know what it was or how to use it. Lily's mother, who was present, hastened to explain that Lily undoubtedly did know what consciousness means even if she did not know that she knew. And her mother, being (like Lily) something of an intellectual, drew a literary parallel: in Molière's play, *Le Bourgeois Gentilhomme*, Monsieur Jourdain was amazed to discover that he had been speaking prose for the last forty years without realizing it was "prose"; likewise Lily had clearly been conscious for the last eight years without realizing she was "conscious." The point of Molière's joke, of course, was that M. Jourdain already knew perfectly well what prose was, but had never dignified it by that title. Her mother's point about Lily was that she already possesses the idea of consciousness, even if she has not yet learned to name it.

Suppose then that, like Socrates cross-questioning Meno's slave boy (whom Socrates showed to have an unacknowledged grasp of Euclidean geometry), I were to have asked Lily some leading questions. Could I have shown that she already has the idea of "having sensations" as a distinct state of mind? Could I in fact have established that she not only shares my conception of consciousness but agrees with most of the other assertions that I made?

There are, I think, strong grounds for believing that I could have gone at least part of the way. They are that no small girl could possibly have failed to notice—as soon as she could notice anything—the distinction between having sensations and not having them as it occurs in her own case. Every day of the year, and often several times within a day, she has lost this state when she falls asleep, and gained it when she wakes. And there is surely no better way of marking the boundaries of an idea than the repeated demonstration of both positive and negative examples: "now you see it," "now you don't."[101]

"Prose" is defined in the dictionary (Oxford Pocket Dictionary) as "unversified language"—that is, it is defined by means of its negation. And when Dr. Johnson was asked by Boswell, "What is poetry?" he replied, "Why Sir, it is much easier to say what it is not. We all know what light is, but it is not easy to tell what it is [except by contrasting it with dark]."[102] If human beings were in a state of having sensations all the time, then the fact of having sensations might be much less striking, just as if the sun always shone, the fact of "daylight" would be much less striking. But as night follows day across the surface of the earth, so the waking state follows the sleeping state across the surface of a child's mind.

So let me begin there and see where the dialogue with Lily takes me. I hope she will forgive me if I conduct the interview in typically bullying Socratic style (although I doubt that I shall be able to take complete control).

NICK   Lily, I want you to think back to when you were asleep, or forward if you like to when you go to sleep tonight. You'd agree, I'm sure, that there's a big difference between being awake and being asleep?

LILY   Of course there is.

NICK   Suppose I were to ask you what "being asleep" is like. Would you mention for instance that when you're asleep your eyes are closed, you don't move, your thoughts have come to a stop, and you no longer sense anything happening to you?

LILY   I probably would.

NICK   In fact it's as if there's a kind of pause in your existence.

LILY   Yes.

NICK   If we wanted an analogy, we might say it's like the flame on a gas lamp being turned right down: it shrinks almost to nothing, though it doesn't actually go out.

LILY   Yes. I kind of collapse into myself.

NICK   Now, if I were to ask you what "being awake" is like, would you say it's just the opposite of being asleep? In other words your eyes are open, you're moving about, and you're having all sorts of thoughts and feelings. As if the flame has come to life again.

LILY   That's right.

NICK   Let's talk about "being awake." What makes it really differ-

ent from sleep? Do you suppose all those things you mentioned are equally important? For instance, when you are
awake do you have to be moving about?

LILY   No, that's not how it is. I usually am moving about, but I
don't have to be. . . . Look, now I've closed my eyes and I'm
not moving at all, but I'm still awake! Once I woke up in the
night after a bad dream, and I couldn't move even though
I wanted to. It was like being paralyzed . . . but I was awake
and I was scared.

NICK   Then, maybe it's your thinking that makes all the difference.
When you are awake do you have to be having thoughts?

LILY   Well, mostly I seem to . . . mostly I'm thinking when I am
awake—even when I'm lying in bed or sitting still.

NICK   I remember a cartoon in a magazine called *Punch*. There was
an old man sitting on a park bench, and there was a lady
who said to him, "Tell me, my good man, how do you pass
your time?" And he said, "Well, ma'am, sometimes I sits and
thinks; and then again, sometimes I just sits." . . . Don't *you*
ever just sit without thinking?

LILY   Well, no, I don't often just sit . . . but sometimes I just lie in
the bath, or just listen to my tapes, or when I'm hurt I just
cry and feel sad . . . or I might just sit and eat an ice cream
. . . and I wouldn't be thinking. Sometimes people say
"Penny for your thoughts," and I don't know what to say
because I wasn't having any thoughts.

NICK   But that doesn't mean you've gone to sleep, does it?

LILY   Of course not.

NICK   So thinking can't be all *that* important to being awake. Then
what about the last thing you mentioned, sensing things
happening to you? When you are awake are you always
having sensations of some kind? Or is it the same as with
thoughts: sometimes you are, sometimes you aren't?

LILY   It depends what you mean by sensations. I'm always having
feelings—when I'm awake, that is.

NICK   Such as?

LILY   Well I'm seeing the blue sky, or hearing a bus go by, or
feeling cold . . . or happy or sad . . . or maybe simply feeling
"here I am."

NICK   Don't all of those involve *sensations:* the impression that
something is happening to you or inside you? You see the

light with your eyes, you hear the sounds at your ears
. . . being happy or sad involves your face, or your limbs,
or your tummy. Even the feeling "here I am" comes down
to something like that. William James—you won't have
heard of him—reckoned that "here I am" meant not much
more than "here I am having these sensations in my head
and neck."

LILY   Yes. But still I'm more used to talking about "feelings" than
"sensations." It's the word I know.

NICK   All right, I don't think we disagree. The point is that, if you
are awake, "feelings"—as you put it—are the one thing you
can't do without. And if someone said "Sometimes, I sits and
has feelings; and then again sometimes I just sits," it
wouldn't make sense?

LILY   I'm not sure about that. Suppose I were to be thinking while
I sit (I agree I don't *have* to be—but suppose I *were*). And
suppose all my other feelings were to stop. Then I'd be sitting
and just thinking—and not necessarily having *any* feelings.

NICK   Well, you *say* that. But do you really believe that's ever quite
what happens? Try it. Close your eyes. I'll count to three.
Then sit and think—and try to block out everything else for
the next ten seconds. One two three. . . . You can open your
eyes now. What was that like?

LILY   My nose was tickling, so it wasn't a fair go.

NICK   Okay. But I think you'll find it never *is* a "fair go": there's
always something that intrudes. All the same, I take your
point. Suppose you could do what you said, and block out
other sensations. Then the question becomes whether thinking itself doesn't have some kind of a "feel" to it.

LILY   You mean, like the man said, sensations in my neck and
head?

NICK   No, actually that's not what I meant (though it's interesting
you should suggest it—there was a whole school of psychologists last century who claimed that thinking does involve feedback from the skin and muscles). What I meant
was that thinking always involves *imagery*, and imagery has
at least a shadowy connection to sensation. Thinking in
words for example is a bit like hearing words, or thinking in
pictures is a bit like seeing them.

LILY   Only a bit.

NICK But enough, maybe—enough for there to be something it *feels like* to think.

LILY Are you saying everything we do involves sensations?

NICK No, just that it's impossible to imagine being awake—or being oneself—without them.

LILY When you put it like that, I suppose you must be right. If I didn't have *any* feelings, it would be the same as if *I* wasn't there.

NICK But where's that leading us? Does it mean that "having feelings" and "being awake" come to the same thing?

LILY Seems like it, though I wouldn't have thought they were quite the same.

NICK Perhaps being awake is more of a *lasting state* that you enter or leave, whereas having feelings is more of a *transient process* that is happening to you right now. You might want to say, for example, "The period when I'm awake is made up of lots of moments of having feelings."

LILY Yes.

NICK But aren't there other reasons too why "having feelings" and being awake don't exactly go together? Perhaps there are times when you have feelings even when you're *not* awake.

LILY Yes, that's what I just remembered. I have feelings—what you'd call sensations—when I'm *dreaming*. When I had that dream I mentioned, I felt all sorts of horrid things were happening to me: I was in the sea, and I was drowning, and I saw the great black monster coming. . . . Usually, though, I have nice dreams.

NICK Someone might say "Sleep well, sweet dreams." They'd be saying "have a good time and feel nice *while* you're asleep," wouldn't they?

LILY Mummy says that.

NICK So I guess it means we need another word for "having feelings." "Awake" won't do.

LILY Yes. And we need a word for "not having feelings," because "asleep" won't do.

NICK How about the words "consciousness" and "unconsciousness"?

LILY But I told you already I don't know what "consciousness" means.

NICK   And I'm saying you *do know* what "consciousness" means. If you're having feelings—or sensations—you're "conscious."

LILY   And if I'm not having them I'm "unconscious"? It must work for cats too, because I heard the vet saying that Prune—that's my cat—wouldn't feel anything during the operation because she'd be "unconscious."

NICK   That's right.

LILY   Well, I've been "conscious"—on and off—for eight years, and I didn't even know it! There's a play by Molière where—

NICK   Lily, you're stealing your mother's lines. . . . Let's leave it there.

LILY   I just wanted to say one more thing. I wonder if it works for Martine—that's my doll. I wonder if she's conscious.

NICK   What do you think?

LILY   No, I don't think so. I mean I don't think Martine ever has had feelings—because *she* doesn't seem to care what happens to her (though I do). But my friend has a walky-talky doll, Amanda, who cries if you pinch her. If Prune is conscious, I wonder if Amanda could be.

NICK   What would it depend on?

LILY   It would depend on whether Amanda actually feels things like I do. I suppose she could do. But I don't believe she does. I think there's a difference between behaving *as if* you're conscious and *actually* being conscious.

NICK   I think so too. But, Lily, you're jumping several chapters.

I do not pretend that conversations even approaching this level of sophistication ever actually take place. But I suggest that something not unlike this process of critical reasoning does go on in every child's mind. By the noting of resemblances and contrasts in her own experience, the child comes to recognize the state of "having sensations" as a natural kind: a state with well-demarcated boundaries, which either exists or does not exist as a fact of life; a fact—on and off—of her own life, and potentially a fact of other creatures' lives.

The subsequent discovery that there is a word in the English language that may name this state is sure to take much longer. In the absence of encounters like the foregoing one, I doubt that any child is ever actually taught how to use the word "consciousness." Instead she has, as it were, to rely on intelligent eavesdropping. She has to note the existence of the word in the speech and writing of others, to get a fix on how those other people use it, and thus eventually to match the word with her preformed idea.

Locke, as so often, put his finger on the problem here: "If we will observe how children learn langauges, we shall find that, to make them understand what the names of simple ideas or substances stand for, people ordinarily show them the thing whereof they would have them have the idea; and then repeat to them the name that stands for it; as white, sweet, milk, sugar, cat, dog. But as for mixed modes [such as consciousness] the sounds are usually learned first; and then, to know what complex ideas they stand for, they are either beholden to the explication of others, or (which happens for the most part) are left to their own observation and industry."[103]

But the process of discovering what the word "consciousness" means is probably never quite finished. And it could be said that you and I—through our own observation and industry—are engaged in it still.

I can do no better, then, than to set out bluntly the results of my own observations of how the term consciousness is used in the linguistic environment from which I come. They are that, whenever "being conscious" as such arises as a topic for discussion, people's primary interest is almost always in sensations: sensations, that is, in the strict sense of affect-laden mental representations of "what is happening to me as an embodied being." And perhaps nine times out of ten the focus is especially upon affect.

Thus, when someone says "the astronauts lost consciousness before the space shuttle hit the sea," the chief implication is: it did not hurt. "The patient was conscious throughout the operation": it did hurt. "You can't deny that chimpanzees are conscious": chimpanzees feel pleasure and pain like we do, and mind about what is done to them. "LSD is a consciousness-expanding drug": it makes a person especially receptive to strange and interesting sensations. "There couldn't be art without consciousness": no one would bother with music or painting unless they were moved aesthetically by their experience. And so on.

In more theoretical discussions too, it almost always comes back to the same thing. "Could a robot controlled by a computer be conscious?"—not unless it experienced colors, pains, itches, etc., and minded about them much as we do. The mere fact that the robot might be able to perceive or to think at a high level would count for nothing if it had no feelings.

This latter is probably the standard popular objection to almost every "scientific" account of consciousness on offer. Indeed, when, in a previous book, I myself proposed that consciousness involves a particular sort of "thinking about one's own mental states," the psychologist Stuart Sutherland replied in a review: "There is, unfortunately, an obvious fallacy in Humphrey's argument. The brain could represent the processes underlying motives, thinking and so on and could use this representation as a model for others' behavior without the representation appearing in consciousness."[104] He was expressing, I presume, the well-worn opinion that consciousness— real consciousness—has to involve the raw feel of "what it is like to be me," and that no kind of abstract computation would be likely to provide this (at least as usually envisaged).

All I can say is that I have myself now moved much closer to this plain man's point of view. I agree that "what it is like to be me" is always as a matter of fact to be experiencing some kind of sensation: indeed that the having of sensations constitutes being conscious, and no human being or animal or robot would or could be conscious without it.

Hence I would agree that any theory of consciousness that is not a theory of the having of sensations has failed to address the *real problem*. But I should reemphasize that I accept this now only because we have discovered (as I doubt that Sutherland had done) a way of reconciling the absolute centrality of sensation with an apparent contradiction: namely that certain states of mind can also enter consciousness that do not arise directly from stimulation of the sense organs. It is even possible that a person might in certain circumstances "just sit and think" and be conscious of his thinking—but only because such conscious thoughts (unlike unconscious thoughts) involve auditory or visual images, and these in turn have a sensory component to them. A robot, by contrast, could perfectly well sit and think without having any such imagery at all.

· 17 ·

FIVE
CHARACTERISTICS
IN SEARCH OF
A THEORY

The *truth or* otherwise of the assertions about consciousness made in Chapter 15 was bound to depend heavily on definition. I would like to think that, in responding to the challenge to bring the word into the open, I have in fact (and hardly by coincidence) provided some justification for most of those assertions—and slipped my hands around the body of consciousness itself. But, with the "real problem" now exposed and delineated, the real work of this book still lies ahead. Indeed everything up to this point might be considered an extended preface to this one question: if to be conscious is essentially to have sensations, then *what is it to have sensations?*

When for example "I have a pain," who or what is the "I" here; in what way does the having of sensations become a property of such an "I"; and how can this "I" with its sensations be set in a material brain? If we can provide answers to these questions, I dare say we shall have consciousness *and* the mind–body problem licked.

The question "what is it to have sensations?" is—or will need to be—a different one from the question of what the functional value of sensations is, or of why sensory representations should play a part in mental life at all. My line has been that the function of sensations is to provide the subject with representations of "what is happening to me"—originally to serve as a mediator of affect, but later with important secondary uses in connection with perception and imagery. These functional ends do not however determine the precise means.

Consider (since for some reason the analogy comes to mind) the case of paying a telephone bill. The payment in question is the transfer of £165 to British Telecom. This is the function that the payment has, and it is what will have been achieved once the payment has been made. But, as the bill states on the back, there are a variety of ways that I could make the payment: by cash, by check, by direct debit, by credit card . . . by mail, at a bank, over the counter. Since they would end up achieving the same thing, the difference between paying by cash and paying by credit card might be said to be incidental or even epiphenomenal. Yet these different ways of paying are of course notably different. If I pay by cash I shall be immediately poorer, but if by credit card I shall remain financially in limbo for a while.

Now, by analogy, although my sensations do indeed have the function of representing what is happening to me, there could in principle be a variety of ways of doing it, and perhaps not all of them would actually be conscious. Thus there might be—and in fact there are—circumstances where I take in and respond to information about what is happening at my body surface without my *feeling* anything at all. The most obvious examples occur during sleep. If my foot is pinched when I am fast asleep, I shall pull it away; or if my eyelids are drawn back and a light is shone in my eye, my pupil will contract: but the plain fact is that I remain unconscious, and in neither case do I feel anything. Given that a human being can respond like this, then so too presumably can other organisms. When an earthworm for example reacts to a pinch or to light arriving at its skin, it need not be any more conscious of the sensation than I am when I am asleep.

With human beings, then, the question must be: what is going on when we form the representations that *are* conscious? How is this representing done, where does it take place, how long does it last, and so on? And since it is our own experience we are talking about, the answers (when they come) had better do justice to our own inner picture of the representing process.

I am going to start therefore by listing some salient introspective observations about what the having of sensations is like. By "salient" I mean both personally salient insofar as they strike me as obvious and interesting, and philosophically salient insofar as they indicate that sensations have some peculiar and rather strange properties (which, among other things, gives them a logical status different from that of perceptions).

Some of this will be old hat. There has been a philosophical tradition of claiming that sensations are special in at least the following respects: sensations are private, intrinsic, ineffable, and directly apprehensible. My own list of special features overlaps with this to some extent: sensations *characteristically* (i) belong to the subject, (ii) are tied to a particular site in his body, (iii) are modality-specific, (iv) are present tense, and moreover (v) are self-characterizing in all these respects. These features—which I shall summarize in a moment and expand on more fully later on—are not necessarily independent of each other. Once we have a decent theory of sensations we may indeed discover that they are all part of the same package.

I have emphasized "characteristically" above because I want to use the word in a very strong sense.

When I say that "Xs characteristically have a property p" I do not mean solely that all Xs as a matter of fact have a property p—as for example all people as a matter of fact have a name. Nor that all Xs necessarily have a property p, as for example every person necessarily has a place where he was born. Rather, I mean that for an X to have this particular property is what makes it this particular X: in other words, an X cannot be individuated or characterized as the X it is, without making reference to p.

In this strong sense I might say, for example, that "coins characteristically have a value," the reason being that a particular coin cannot be characterized as the coin it is without mentioning how much it is worth; or that "countries characteristically have boundaries," the reason being that a particular country cannot be characterized as the country it is without mentioning where the boundaries run.

What it means to say that something is "self-characterizing" is less easy to exemplify—not least because there are very few things other than sensations that are self-characterizing in the sense that I intend. But when I say that "X is self-characterizing as having property p" I mean roughly speaking (for the moment) that X "tells its own story" in such a way that anyone who is in the presence of X is immediately and automatically aware that X is p. This amounts of course to more than merely being characteristically p: the fact that coins characteristically have a value does not imply that anyone who holds a coin in his hand immediately knows this value, or the fact that countries characteristically have boundaries does not imply that someone living in a country immediately knows where the boundaries run. But the fact that sensations are self-characterizing *would* imply that anyone who feels a sensation immediately knows what its properties are.

1. SENSATIONS CHARACTERISTICALLY BELONG TO THE SUBJECT

The starting point for all of this is that "what is happening to me" is what is happening to "my *embodied* self." The body of every individual human being—contained within the physical membrane that marks the physical boundary between "me" and "not-me"—is structurally, physiologically, and in many respects informationally isolated from every other body in the world. What happens to this particular body is of primary and paramount concern only to the living person that it houses. The very word "life" comes from *leib*, meaning body, and it is no accident that we use the words some*body* or any*body* as synonyms for some *person* or any *person*: to have a different body is to be a different person with a different life.

Hence a sensation that represents "what is happening to my embodied self" obviously cannot be characterized as the sensation that it is without mentioning which body it relates to. It is not just that the sensations I feel happen, incidentally as it were, to be associated with this body. It is that if they were associated with any other body they would have to be different sensations. When I feel a pain in my toe, I feel it in *my* toe, and no description that failed to mention that the toe was mine would be complete.

The sensations I feel are therefore inalienably mine: I have a proprietorial relationship to them—I own them—in a way nobody else does or could do. The pain in my toe belongs to me, and could not even in principle be shared with or transferred to anybody else.

It is true that I and somebody else could each feel very "similar" sensations. When for example we both look at the same rainbow, taste the same asparagus, or hear the same opening bars of Beethoven's Fifth Symphony, we probably do feel sensations that are very much alike, since in such circumstances what is happening to my body is bound to be very similar to what is happening to the other person's. Yet the crucial fact remains that "what is happening to me" is happening to *me* and "what is happening to him" is happening to *him,* and insofar as I and he are separate beings these sensations can never be the same.

There would, of course, be no bar in principle to somebody else getting to know about what is happening to me through other means than by having a sensation of his own. For "what is happening to me" could, in certain circumstances, be for him a part of "what is happening out there": in other words he could *perceive* the same events at my body surface that I myself am *sensing.* He might for example see with his eyes that there is a thorn in my foot, he might feel with his hands that my brow is hot, or he might hear with his ears that I am sneezing. Yet even though he might thereby get to know the same objective facts, he would not be experiencing them as I am.

Since I too can take a third-person view of my own body, it is not only someone else but I too who can perceive what is happening to me as a special case of what is happening out there. And it is important to note that my perceptions of my body, unlike my sensations, are not privately owned in this way. If I run my fingers over a bruise on my forehead, for example, I can perceive that there is a lump under the skin; and if you were to run your fingers over the bruise you could perceive the very same thing. But the difference between us would be that when I run my fingers over the bruise I have both a perception that there is a lump and a sensation of the lump being touched, whereas when you run your fingers over the bruise you have the same perception but miss out on the sensation.

Perceptions in general are not privately owned, because "out there" is generically distinct from "me, my body." Hence a perception of what is happening out there can usually be characterized as the perception it is without mentioning the subject or his body in any way at all. When, for example, I have the perception that "there is a red apple on the table" or that "the clock is chiming four," the content of these perceptions has nothing especially to do with me. Equally, when I and someone else see the same rainbow, taste the same asparagus, or hear the same music, there is no reason why the content of our perceptions—as distinct from our sensations—should not be virtually identical.

"Many people," as Milan Kundera has written, "few ideas: we all think more or less the same, and we exchange, borrow, steal thoughts from one another. However, when someone steps on my foot, only I feel the pain."[105]

2. SENSATIONS ARE CHARACTERISTICALLY TIED TO A LOCATION IN BODILY SPACE

There is more to the bodiliness of sensations than just the fact that they belong to one person rather than another. For, besides occurring in my body, my sensations always occur *somewhere in particular*. What matters of course is not the absolute location defined in physical space, but rather the location defined according to bodily coordinates: whereabouts in my body space it is located. If I touch a stinging nettle with my foot, and then touch the same nettle with my hand, I have two different sensations, even though the event that gives rise to them may have occurred at the same physical location.

Hence a sensation cannot be characterized as the sensation that it is without mentioning *where* in this body space it is occurring. When I feel a pain in my toe I feel it in my *toe*, and no description that failed to mention the toe would be complete. It is not just that the sensations I feel happen to be located at the sites they are: it is that if a sensation had any other location it would be a different sensation—a pain in my toe is a different sensation from a pain in my thumb.

This feature of sensations is probably most obvious for the sense of touch, but it holds equally for the other senses. My sensations of taste have a felt location in the region of my tongue, my sensations of smell in the region of my nostrils. Likewise, my sensations of light and sound have a felt location in my visual and auditory fields. With tastes and smells the location may not be all that precise; nonetheless a sensation of sweetness at the tip of my tongue is a different sensation from a sensation of sweetness at the back of my tongue, and neither could possibly be mistaken for a sensation of sweetness in my knee. With lights and sounds the location in the field is considerably more precise: so that, for example, two stars separated by just a few degrees of angle in the visual field give rise to quite distinct sensations, as do two clicks separated by a few degrees of angle in the auditory field.

Admittedly, in the case of the visual and auditory fields the sensations are not actually felt to be located at the body surface—in the retina of the eye or the basilar membrane of the ear as such. Instead these fields are constituted by a set of radii centered on the head, defining a kind of visual or auditory capsule. They are nonetheless part of my body space and move with my eyes or my head. If I form an afterimage of a bright lamp in the dark and then move around, the sensation stays in the same place in the visual field and moves around with *me*.

There is again a clear contrast between sensation and perception here. My perceptions need make no mention of my body, and it follows, *a fortiori*, that they need make no mention of any particular region of my body space. This is true even though the perception may of course concern a particular outside location. When I perceive with my right hand that "there is a nail in such and such a position in the floorboards," the perception could perfectly well be characterized without mentioning my right hand, and indeed I could have had exactly the same perception using my left foot (while having a different sensation). Equally, when I perceive out of the corner of my eye that a bird has just alighted on the windowsill, the perception could be characterized without mentioning which corner of my visual field was used, and indeed I could have perceived exactly the same thing out of the other corner of my eye (while again having a different sensation).

3. Sensations Are Characteristically Modality-Specific

There is more to it still. Besides having a particular location my sensations always belong to a particular qualitative category, related to *what kind of thing* is happening to me—whether the stimulus at my body surface has the form of mechanical pressure, heat, light, sound, aroma, or whatever—and how in particular it is affecting me.

Thus every sensation that I feel belongs to a distinctive "sensory modality," tactile, visual, auditory, olfactory, gustatory, or a submodality of one of these. A sensation cannot be characterized as the sensation that it is without mentioning *which sensory modality* it belongs to. When I feel a pain in my toe I feel it as a *pain*, and no description that failed to mention its painfulness would be complete.

Again, it is not just that the sensation happens to have this quality associated with it: it is that if the sensation had a different quality, it would be a different sensation—a sensation of tickle on my tongue is a distinctly different sensation from a sensation of sweetness, even though it may occur in the same place.

There might seem to be an evident connection between this feature of sensations and the previous one: between sensations having a modality and having a definite location in body-space. There is certainly a remarkable correlation between the two, for it is a fact of life that taste sensations occur only in the mouth, visual sensations in the eyes, and so on. But this correlation between location and modality is presumably in part an accidental one—a consequence of the way human bodies happen to be built. Although I have never had taste sensations in my ears, or visual sensations in my nostrils, I can conceive that if I were a different sort of creature I might do so. Just as I myself can have both tactile and gustatory sensations in my mouth, if I were an octopus I might have both tactile and gustatory sensations in my arms.

I shall say more, later, about the nature of sensory modalities. Their absolute distinctiveness—the gulf between one modality and another—is one of the most mysterious facts about sensations. Each modality is as it were a separate territory, within which (at least in imagination) it is possible to travel smoothly, but between which there is no imaginable bridge. I can imagine moving through an unbroken line of intermediate sensations from red to green, sour to sweet, tickle to itch, C sharp to A flat, but by no stretch of the imagination can I get from red to sour, or tickle to A flat.

This gulf between sensations of different modalities is certainly more absolute than that which exists between sensations at different locations. I can imagine a continuous line of pain sensations from my tooth to my cheek to my eyes, and I can even imagine (if I try really hard) a continuous line of visual sensations from my eyes to my tongue. But what I simply cannot imagine is a tactile sensation at my tongue becoming by continuous progression a visual one: of that it seems as hard to conceive as of a tactile sensation at my tongue becoming a tactile sensation at somebody else's tongue—almost as if the different modalities involve two separate owners.

However this may be, let us note again how sensations, by being modality-specific, contrast with perceptions. Since perceptions are concerned not with the nature of the stimulus as such but rather with

what it signifies in the external world, they need make no reference
to a sensory modality at all, and are in fact essentially amodal.
Indeed there is no reason in principle why one and the same perception should not be mediated by completely different sensory systems.
I might for example arrive at a perceptual representation that "it is
raining," or "there is a dog at the door" by means of my eyes, ears,
skin, nose, or a combination of all four. Moreover, in the peculiar
case of "skin-vision" discussed earlier, we have a case where someone can have a typically visual perception, for example that "the
moon is up," or "there is a triangular object in the corner of the
room," by making use not of his eyes but of the skin of his back.

4. Sensations Are Characteristically Present-Tense, Existing Entities

A further fact that results from sensations being representations of
"what is happening to me" is that sensations have a *time* that they
refer to: namely the time when what is happening *is* happening, the
"present." All sensations are, strictly speaking, present tense. When
I feel a pain in my toe, I am feeling that there is a pain right now;
and it would make no sense for me to be feeling that there was a pain
yesterday, or that there will be one tomorrow.

Moreover, sensations have a significant "lifetime." That is to say
every sensation persists for roughly so long as the surface stimulation continues. Although the lifetime may be very brief, as with the
sensation created by a flash of lightning, even so the sensation must
last for at least a moment before ceasing. It follows that sensations
can be said to *exist*, and even to exist as *individual entities*. When I
feel the pain in my toe, the sensation starts at a certain time, lasts
so long, and eventually dies away or changes. But, while it lasts, it
is the same individual pain. And if, after stopping, the sensation were
to start up again, this would now be not the earlier pain but a new
pain of the same sort. Equally, when I look at the green walls of my
study, I feel a green sensation that remains the same sensation until
I look away. And if, having looked away, I look back, the green
sensation I now feel is not the same but a new instance of the earlier
one.

Hence every sensation is necessarily in existence at the time I feel it. And a sensation cannot be characterized as the sensation it is without mentioning when this present time is. It is not just that it happens to be occurring now: it is that if the sensation occurred at any other time it would be a different entity.

As it happens, at any one time we are always the subject of a whole population of existing sensations that have lasted more or less long. At this moment for example I have been feeling cold for several minutes, smelling the coffee aroma for about thirty seconds, and seeing and hearing my current visual and auditory sensations for a variety of periods down to a mere fraction of a second. All these coexisting sensations collectively make up what is currently "in consciousness," and together they might be said to constitute the "conscious present."

In all these respects sensations differ from perceptions. For a start, perceptions can refer not only to the present but also back to the past or forward to the future. We can perceive not only that it is raining, but that it has rained or that it will rain. But furthermore, perceptions unlike sensations do not *exist* for any length of time. It may, it is true, take us some time to take in the information required to arrive at a perception. But the perception itself is not an enduring entity with a life of its own. In fact, in terms of grammar, perceptions are properly speaking always "perfective"—already complete—whereas sensations are generally "imperfective"—continuing and unfinished. "I perceive the traffic light is red" implies I have just now, but already in the past, perceived it; whereas "I sense a red sensation" implies I am still, in the present, sensing it.

5. SENSATIONS ARE SELF-CHARACTERIZING IN RESPECT OF PROPERTIES 1 TO 4

Now we come to perhaps the most fundamental—and most perplexing—of the features that I listed, which is that sensations are self-characterizing or self-disclosing. Sensations tell their own story or give away their characteristic properties, so that the subject is directly and immediately aware of them.

When I feel a pain in my toe, the sensation is there for me as the sensation that it is, without my having to do any kind of mental work to classify it as being one sensation rather than another. Indeed my impression in this case is simply that *my toe hurts*: and, when my toe hurts, the facts that it is my toe (rather than some other bit of me), that it is acting in a painful fashion (rather than acting in a visual, gustatory, or auditory fashion), and that it is hurting now (rather than some other time) are, as it were, primitive facts, about which I could not possibly have any doubt. I certainly do not have to "work out by inference" that it is "probably" my toe rather than yours, my toe rather than my thumb, a pain rather than a smell, present at this moment rather than five minutes ago. Instead it seems that these properties are implicit in the sensation, so that probability and inference do not come into it. The sensation is, if you like, "phenomenally immediate."

One of the striking consequences of this, which also brings home the reality of the phenomenon, is that I can feel the sensation aroused by a stimulus before I am in any position to analyze the stimulus in terms of what it signifies, let alone to describe it in words: my sensations include at any one time much more than I have yet come to terms with at a *perceptual* level. While this is true for all the sensory modalities, it is perhaps most obvious for vision. When I am in a dark room and the lights are suddenly turned on, I all at once experience color sensations throughout my visual field (even if they are a bit fuzzy and washed out toward the edges). Yet, while I am feeling this full field of sensations, I am at first far from being fully perceptually informed about the room. In fact when the light comes on and I take in all the colored patches as sensations, I am at first in the peculiar position of, as it were, "seeing beyond my means"—I am feeling sensations that I have as yet no way of paying for in terms of categorical description.

The point can be illustrated in a more elementary way by a reaction-time experiment. Suppose a light of one of several colors appears on a screen in front of me and I have the task of identifying the external color as fast as I can—that is, identifying it perceptually—and pressing one of a set of corresponding buttons. Then if there were a choice of only two colors—red and green—and two buttons, it might take me about a quarter of a second to respond. But if there were a choice of eight colors—red, orange, yellow, blue, green, white, pink, and violet—and eight buttons, it would probably take me nearer a second to respond. The reason is that in the former case I am having to make only one binary decision, but in the latter case three; and every decision made at the perceptual level takes an appreciable time. Nonetheless, even though it takes nearly a second to *decide* that one of eight colors is yellow, it does not take anything like that long to *feel* the yellow sensation. In fact I would say I feel the sensation almost instantaneously no matter how many alternatives there are to choose from—and indeed that I feel it without making any decision whatsoever.

How this can be, and what it means, will present major problems for a theory of sensations. But here is a first thought as to the answer. Taking again the example of my toe, my impression, as I said, is that when I feel the pain my toe hurts. But there is more to it than this. For, if my toe is actively hurting, and my toe is part of me, then perhaps it makes sense to suppose that at some level I am actively involved in doing the hurting. Indeed, rather than simply receiving the sensation, it could be that I am actively creating it, even *issuing instructions for* it—so that feeling the sensation has something in common with an *intentional activity*. And, if this is how it is, then the instructions I am issuing to bring this particular sensation into being would be the primary fact before my mind. Hence I would no more need to "ask myself what I am doing" when my toe hurts (or my eyes sense yellow) than I would need to ask myself what I am doing when I instruct my arm to wave.

I could list other features of sensations besides these five. But these must be enough to be going on with. If we can provide an answer to how these five features of sensations could emerge as logical/ biological corollaries of a plausible mechanism in the human brain, we shall be doing better than any theorist has so far.

The search for this answer begins in the next chapter.

A cautionary tale may be in order. As a boy I went fishing on one of the Norfolk Broads and hooked a large pike, weighing twenty-three pounds. I fought it for nearly an hour before dragging it on shore. I knocked it on the head, bundled it into a sack, strung it under the crossbar of my bicycle, and rode the five miles to my grandmother's house. The advice in Mrs. Beeton's cookbook was to soak the pike in salt water for twelve hours. I filled the bathtub with water, poured in a box of salt, and heaved my dead pike in. A few hours later, when I was reading by the fire downstairs, I heard a great splash. The pike had come to life again, had leaped out of the bath and was thrashing about on the floor. The moral of which is that it is one thing to catch a fish but another to prepare it for the pot.

18

THE PROBLEM OF OWNERSHIP (A TACK TO STARBOARD)

When *I feel a pain* or a taste or a sensation of red light, the experiences belong exclusively to *me*, they are *my own*.

This was cited above as the first—perhaps most obvious—feature of sensations, and both its truth and its meaning were assumed to be intuitively clear. Yet the idea of "ownership"—especially inalienable or private ownership—is, when you consider it, a very strange idea indeed.

There are more riches here than have yet been revealed. But to reach them the discussion will have to range more broadly. For it is not only in relation to sensations that problems may arise as to precisely what "ownership" can mean. Here are some other things that are my own: my house, my garden, my bicycle, my dog, my shoes, my feet, my voice, my memories, my deeds, my image in the mirror, my writing of this book. And if some of these examples seem a little puzzling, consider the still more remarkable claim to ownership made by the seventeenth-century English mystic Thomas Traherne: "The streets were mine, the temple was mine, the people were mine . . . the skies were mine, and so were the sun and moon and stars; and all the World was mine, and I the only spectator and enjoyer of it."[106]

Even with external objects (which most people would probably take to be the paradigm example) the nature of the relationship between the owner and the thing owned is far from being theoretically transparent. I say the garden is mine: it belongs to me, I own it. But how would I explain this to someone who did not already know what I was talking about? Jean-Jacques Rousseau, in the *Discourse on Inequality*, wrote: "The first man who, having enclosed a bit of land, said 'This is mine,' and found people stupid enough to believe him, was the true founder of civil society."[107] But perhaps it was not a matter of whether they believed him, but rather of whether they understood just what he meant.

The linguist Ray Jackendoff asks in a recent paper: "What does X owns Y mean?" And answers his question: "Very roughly there seem to be three parts: (a) X has the right to use Y as he wishes. (b) X has the right to control anyone else's use of Y, and to impose sanctions for uses other than those he permits. (c) X has the right to give away rights a and b."[108] Thus Jackendoff, like Rousseau, takes ownership to be essentially a *social* concept, based on an agreement by others that the owner has certain special rights. In fact he goes on to suggest that the concept of ownership, along with others such as kinship and dominance, may actually be innately prefigured in the human brain as part of a "social cognition module" that has evolved in the later stages of primate evolution—a sort of innate social grammar.

There is much to be said for the idea of a biologically based social grammar (and in a different context I have made a similar suggestion[109]). But I do not think that ownership belongs there, or at least that it originates there. For, if the concept of ownership is essentially social, it would imply that the concept could not have arisen until people had an understanding of social rights. And this seems quite unlikely. Even if it is by agreed right that a person owns his worldly goods, or even if it is by agreed right that a person owns his shoes, it can hardly be by agreed right that a person owns his feet. And if someone owns his feet, and he himself knows that he owns them, then it seems very probable that he has—and always would have had throughout human history—the basis for understanding ownership in general.

It is arguable—and I think true—that the whole idea of so-called private property is psychologically nothing other than a metaphorical extension of the idea of "my body, my self"—a matter of setting the boundaries that much wider. People (and not just people: watch a dog with a bone) certainly behave as if they regard an intrusion on or insult to their private property as tantamount to a threat to their bodily well-being. Steal someone's goods and he may feel personally violated; trespass on your neighbor's plot, and he may assume as much right to expel you as if you had trodden on his toe. "Our bodies are our gardens," Iago said. And our gardens, our cars, even our money in the bank, are often treated as outposts of our bodies. It is even so with a man's works: see how an author reacts when someone else steals his ideas.

Let us suppose then that the idea of ownership began (and still begins in each of us) not as a social concept at all, but as a highly individualistic one. Rather than external objects providing the primary example of ownership it is actually the other way around. "Mine" gets its meaning from "me." The first things that belong to me are those things that are actually physically a part of me—and it is only later that the concept gets extended to other kinds of property.

Yet this will only displace the problem of the origin of ownership without resolving it. For, however primitive and individualistic the idea of ownership may be, we should not imagine that human beings are *born* with the idea of their bodies being their own. Instead, when a baby first enters the world, the physical extent and limits of his own body are presumably things he has to discover by experience: even the ownership of his feet can hardly be a *given*.

So the question becomes how this primary instance of ownership itself takes root. What are the psychological or logical criteria by which an individual establishes—to begin with—that the parts of his own body do in fact belong to him? Is there something that comes still prior to the ownership of bodies, an even more basic instance of belonging, that serves as the ultimate determinant of what else is or is not "mine"?

I think there is, and that it lies with the idea of "I," the owner, being what might be called my "executive self." The central fact of my individual existence as an owner is that "I" am a voluntary agent which has my body under its control.

It would seem to be an analytic truth—and *not* something that has to be established by experience—that the one class of things that "I" as a voluntary agent indubitably own are my volitions: the plans and intentions that originate within my mind and which translated into action constitute the things "I" do. When, for example, "I" will my arm to move, the instruction to move it cannot be anything else but "my" instruction. If such instructions are necessarily mine, it follows that the actions that result from them are also necessarily mine. But, since such actions are as a matter of fact always effected via a particular set of bodily appendages, it then follows as a contingent truth that these bodily appendages themselves are also mine. Moreover since, as a matter of fact, I am alone in having this particular relation to this body, not only is my body mine but in this respect it is privately and inalienably mine.

An unusual case should be sufficiently exceptional to prove the rule: the case of Siamese twins.

Suppose I myself were to have a twin brother, joined at the waist, sharing the same skin with me and some of my internal organs, but each of us with our own (sic) head and limbs. As we know from real examples of Siamese twins, each twin would in fact typically present himself as a separate "I"—a separate agency—who speaks with a separate voice and has his own thoughts, feelings, and so on. Even in law each twin would be regarded as a separate person, and have a right to the individual ownership of property (the twelfth-century pair of female twins, the Maids of Biddenden, had separate husbands and separate sets of children, and before they died they made separate wills). External property apart, however, the first fact is that each twin would confidently claim that certain parts of the joint body were *his* and *not* his brother's.

So, which bits of our joint body would I in such circumstances claim as belonging especially to *me*? What I imagine I would claim as mine, and what real twins do in fact claim, would be the set of limbs that "I" control and speak for. This arm would be mine because it obeys none other than *my will,* that arm would be his because it obeys none other than *his.*

There are plenty of more ordinary situations to confirm that this analysis is valid. In a supermarket, for example, I catch sight of a figure on the security television monitor which bears a passing resemblance to myself. How do I find out if the figure I am looking at is mine? I wave my arm: and if it is my body, it waves back. Or (somewhat more farfetched) I have one of my hands intertwined with someone else's hand and, on looking at this mess of fingers, I am not sure which are mine and which are his. How do I decide about *this* finger? I try to wiggle it: and if it is mine the finger moves.

With adults such "self-tests" are of course usually no more than self-confirming, rather than self-creating and defining. But they play a much more crucial role in early infancy. Human babies (and the babies of many another species too) will be observed to spend considerable periods just watching their own arms and legs flailing in the air—as they investigate through their actions precisely which bits of the world do and do not belong to them. The principle may not be totally reliable, but in the long run it succeeds: "If something moves as and when I will its movement, it is me and it is mine."

In line with this, Daniel Stern, the child psychologist, has described a test he made with two real Siamese twins.[110] These four-month-old girls, Alice and Betty, were connected front to front, at the level of their tummies, so that they always faced one another. Very frequently one would end up sucking on the other's fingers, and vice versa. Assuming the twin who was sucking enjoyed the activity and would want it to continue, Stern's question was: would she know how to respond if the arm was pulled away? Would she know whose fingers she was sucking?

Stern did the following experiment. When Alice was sucking either her own or Betty's fingers, he gently pulled the arm away from her mouth and observed what happened. He found that, if it was Alice's fingers in Alice's mouth, Alice's arm resisted; while, if it was Betty's fingers in Alice's mouth, Betty's arm did not resist and nor did Alice tense her (free) arms—though in the latter case Alice did try to follow the fingers with her head. It seems that Alice undoubtedly knew which parts of their accidentally conjoined bodies were under *her* control. "Alice," Stern writes, "seemed . . . to have no confusion about whose fingers *belonged* to whom" (my emphasis).

What would happen if someone were *not* to have control of their own body? We all know the peculiar experience of having an arm or a leg temporarily "go to sleep" as the result of the blood supply being reduced: just for the moment, the paralyzed limb becomes a kind of alien thing. But if the paralysis were to be much more long-lasting as a result of brain damage the effects might well be more disconcerting still. Do such brain-damaged patients ever *disown* their own limbs?

The answer is that this does sometimes occur (although by no means always). Patients have been described who, when they are paralyzed down one side, flatly deny that the affected limbs belong to them at all.

This is how the neurologist Eduardo Bisiach reports it. "A minimal form of these disorders may be seen in the feeling of extraneity of the limbs, explicitly referred to by the patient, or inferred from the peculiar nicknames they apply to them. . . . In the severe form the patient maintains that the limbs belong to someone else, e.g. to the examiner. The content of the delusional beliefs may be utterly absurd: the patient may claim that the arm belongs to a fellow patient previously transported in the ambulance, or that it has been forgotten in the bed by the previous patient. Sometimes the patients have a quite tolerant attitude to the repudiated limbs, while in other instances they are irritated by their presence and insist on having them taken away. In some cases, albeit infrequent, states of furious hatred towards the alien limbs, and even physical violence may be observed."[11]

Bisiach relates the following interview with a patient who was paralyzed down the left side of his body (and also blind on that side):

The examiner, placing the patient's left hand in the patient's right visual field, asks "Whose hand is this?"

PATIENT: Your hand.

The examiner then places the patient's left hand between his own hands, and asks: "Whose hands are these?"

PATIENT: Your hands.
EXAMINER: How many of them?
PATIENT: Three.
EXAMINER: Ever seen a man with *three* hands?
PATIENT: A hand is the extremity of an arm. Since you have three arms it follows you must have three hands.

The examiner then places his hand in the patient's right visual field, and says: "Put your left hand against mine."

PATIENT: Here you are [without performing any movement].
EXAMINER: But I don't see it and *you* don't see it either.

PATIENT: [After prolonged hesitation] You see, doctor, the fact that the hand didn't move might mean that *I* don't want to raise it . . .[112]

Thus the patient not only denies that the hand belongs to him, but when challenged by the circumstantial evidence ends up by raising doubts about his own intentions—doubts that are almost certainly not wholly sincere, for we can almost hear him whispering to himself, like Galileo, "But I did want to move it." There could hardly be a stronger demonstration of the link between the self as owner and the self as agent.

"Our bodies are our gardens," Iago said, "to the which our wills are gardeners."[113]

The starting question was: what does it mean to say that "this is mine," specifically in relation to sensations but more generally in relation to our bodies and the world beyond?

Insofar as "I" am a voluntary agent my *volitions* are my own, and in the normal course of events these volitions specifically and uniquely bring about movements of *my* body. Hence people will consider voluntary control over their own bodies to be the criterion of whether or not those bodies do indeed belong to them. But furthermore, although there is nothing in the outside world that "I" control in quite the way that I control my body, there are other things of which I am the *de facto* controller. Hence, by extension, people will make this the criterion for what else, in the outside world, belongs to them as well.

So we see how Jackendoff's criterion, that for "X to own Y" is for "X to have the right to use Y as he wishes"—or something very like it—might have evolved from bodily beginnings to cover private property in general. Just as my body is mine because I have a natural capacity to do things with my arms, legs, tongue, etc., so my garden, my bicycle, my dog, and even my work on this book are mine because I have the capacity (and a social right) to do things with them.

Indeed it is precisely because *this* is the meaning of ownership that Thomas Traherne's claim that "the sun and moon and stars" are his strikes us as so odd and ultimately silly. For there is nothing that he or anyone else could *do* with the sun the moon and the stars. His horse might belong to Traherne, the crown jewels might, the Taj Mahal might—but not the stars: even Rousseau's noble savages could never be stupid enough to believe that.

And yet Thomas Traherne could *look* at the stars.

> *Look at the stars! Look, look up at the skies!*
> *O look at all the fire-folk sitting in the air!*[114]

He could respond to the light falling on his own eyes, and think to himself: this is happening to me, I am sensing the stars, I am the "only spectator and enjoyer" of this *sensation*.

What, then, of sensations? Could they be mine for anything like the same reasons that my garden, shoes, feet, actions, or volitions are? And, if so, which of these levels provides the proper parallel? Could my sensations be mine because they too—in some peculiar way—are under my executive control?

The way the argument is currently going may not seem promising. (1) My body is mine by virtue of the fact that I can do things with it. (2) My goods, land, etc., are mine by virtue of the fact that I can do things with those too. (3) Conclusion: my sensations are mine by virtue of the fact that I can do things with them too (??).

If this were really the structure of the argument, it would not wash. Nobody does things *with* sensations. While I can wiggle my toes, or spend my money, or fence off my land, I cannot do anything comparable with my pains or tastes or sensations of red light. Sensations are just not the right kind of entity to be made the object of an action in this way.

Then what kind of entity are sensations, and how is it that they are in fact so evidently "mine"? Is it possible that sensations, rather than being objects of actions, are in fact close to being a kind of bodily action in their own right?

Consider for example the grammar of the sentence: "I feel a pain in my toe." The obvious way of parsing this sentence would be "I [subject]/feel [verb]/a pain in my toe [object]," on the model of "I / dig / my garden." But maybe the correct—though not so obvious—way of parsing it would be: "I [subject] / feel-a-pain-in-my-toe [verb]," on the model of "I / wave-of-my-arm." Then the pain-in-my-toe would be a *way* of feeling, not an object of it, just as the wave-of-my-arm is a *way* of acting, not an object of it.

The experience of feeling-a-pain-in-my-toe cannot of course be the same sort of activity as that of waving-my-arm. Pains and other sensations might however be "quasi-bodily activities" which *implicitly* involve some sort of movement in the region where the sensation is being felt—and this would make them at least logically from the same stable as overt activities. Indeed "I," my sensory self, would be in reality but another side of "I," my executive self. "I" would be doing the acting and speaking for my self, and in the end "I" would be doing the feeling too.

There is a lot contained in the preceding paragraphs that—if it does not make immediate sense—will become much clearer shortly. But, as a taste of what is to come, let me round off this discussion of ownership by trying a curious argument.

Consider again the example of my fingers being intertwined with someone else's. If I am in doubt about whether a particular finger belongs to me, I could, as I said, decide the matter by attempting to move the finger voluntarily and observing the result: if it moves when I will it, that makes it mine. But there is an alternative method I could use: I could simply reach over with my other hand and pinch the finger, and if I feel a sensation of pain, that too makes the finger mine.

Now suppose there were reason to believe—I am not yet saying that there is, but I am not saying that there isn't—that the first of these two methods is logically primary, so that in the final analysis the *only* way I could get to know for sure that the finger is my own would be by performing some sort of intentional action with it. The implication would be that my feeling a sensation in my finger would also logically have to involve my performing—or at least intending to perform—such an action.

This is probably too curious an argument to be convincing while it stands alone. But if space can be made for it, then watch that space.

19

THE QUESTION
OF INDEXICALS
(A TACK TO PORT)

At first blush, the idea that sensations are equivalent to bodily activities may sound extremely odd (although readers who have come across the so-called adverbial theory of sensations may not find it as odd as some[115]). Indeed you may be thinking that at best it will provide an interesting analogy but not a theory of what sensations amount to in real terms.

True, the analogy, once one is alerted to it, begins to look surprisingly interesting. For there are certainly formal similarities between the two classes of phenomena, besides the one already singled out. Compare, for example, what it is like "to wiggle my toe" with what it is like "to feel a pain in my toe." In addition to being *mine*, the activity of "my toe-wiggling" resembles the sensation of "my-toe-hurting" in all the following respects.

The activity like the sensation implicates a *particular part* of my body (it cannot be characterized as the activity it is without mentioning where it is occurring—that it is the toe rather than, say, the hand).

The activity like the sensation is a *present-tense process* with its own lifetime (it cannot be characterized as the activity it is without mentioning when it is occurring—that it is this very moment's bout of wiggling rather than, say, yesterday's).

The activity like the sensation has a *qualitative dimension* to it, akin in some ways to having a modality (it cannot be characterized as the activity it is without mentioning the manner or adverbial style in which the bodily movement is occurring—that it is being done in a wiggling manner rather than, say, a grasping one).

Furthermore the activity like the sensation is *phenomenally immediate* (its characteristics cannot but be known to me directly—since I myself, the author of the movement, am actually issuing the instructions for my toe to wiggle).

Yet resemblances, merely at this formal level, are not all that are required for a good theory. And, to make headway in the more ambitious direction proposed in the last chapter, we need to establish that the analogy is actually much closer to a genuine homology: in other words that sensations actually *are* a kind of bodily activity.

Then suppose it could be shown that, beyond these mere resemblances, sensations and bodily activities share at least one crucial property that *only* a bodily activity could have. Suppose we could mount an argument along the following lines: "*Only* bodily activities can have such and such a property; sensations have this property; therefore sensations *must be* a kind of bodily activity."

As it happens, the argument which ended the last chapter had more or less this structure—the crucial property being that of "belonging to me." Thus: "the only way I can establish that a bit of my body belongs to me is through attempting to move it; I can establish ownership over my body by feeling sensations; therefore sensations must involve some kind of bodily movement."

However, while I believe that an argument based around ownership might—with some additional special pleading—be made to work, I reckon it will be more persuasive if it is related to one of the other properties that sensations and bodily activities have in common. And the property that bids to be most promising is the property of being "self-characterizing in respect of its location." Thus what we should attempt to show is that nothing other than a bodily activity can immediately reveal to me (the subject of it) that it involves *this* part of *me*, right *here*.

The key to the argument lies in the words "me," "this," and "here." But in order to develop it, I shall need, as with ownership, to broaden the discussion.

The point has already been made that when I feel a sensation or undertake a bodily activity, then these events cannot be characterized as the events they are without "mentioning" whereabouts in the body they are happening. The question that has, however, not been raised is that of *who* is doing the "mentioning" to *whom*. It might have been assumed all along that "I," the owner of the body, am obviously mentioning the whereabouts to myself. Fair enough. But in that case there are further questions waiting.

When I have a pain in my toe or wiggle it, it is indeed I—the owner of the toe—who seems best placed to mention which bit of my body is being referred to. And it is certainly I who has the primary interest in it and for whom the state first exists as a toe-implicating state. Nevertheless, I could also in most circumstances mention it to someone else: "Where does it hurt?" "In my toe, this very toe." "Which bit is wiggling?" Again "My toe." But then what exactly is involved in mentioning "my toe" to myself—and, in particular, how does mentioning it to myself compare with mentioning it to someone else?

To take the last point first, consider the linguistic substitutions I could make in mentioning my toe to myself. I could say to myself "my left big toe," or I could say "this toe" or "this bit of me" or simply "here"—and in every case I would know exactly what I meant. But if I were to mention my toe in these different ways to someone else, the words "this" or "here" would mean nothing in particular unless I accompanied them by an overt act of *pointing out* the toe; and even if I were to point it out, the words would only make sense to someone else provided he was in my presence and could observe what I was doing. Over the telephone, for example, they would have no role at all!

The words "this" and "here" belong to a class that philosophers call indexicals. The term "indexical" comes from "indicate," and indexicals typically do involve an additional, often nonverbal, act of indicating by the person speaking them. Others of the class are the words "now" and "today," and also the words "I" and "me." All such words get at least part of their meaning from the context in which they are spoken (where, when, by whom, with what accompanying action?).

Imagine, for example, the following exchange recorded on a telephone answering machine. "This is the doctor's office. Please say who you are and when you called, and then mention where it hurts." "Hello, it's me. The date is today and the time is now. The pain is in this bit of my body, right here." While this message might mean everything to the patient, it would convey almost nothing to the doctor.

But for a person to indicate a thing to someone else, what exactly does he have to do? Does he actually have to point to it with his hand (maybe with his "index" finger)? No, clearly not. When I say "this" (meaning for example "this apple on my desk"), I could indicate the object in question by pointing to it, picking it up, throwing it over to you, or sticking a pin in it. Or I could, if I chose, do something rather more elaborate: I could draw a plan of my desk and stick a pin in the plan or write "X marks the spot." But whatever I did, I would have to create some sort of *physical disturbance at a relevant space-time location*—either where the apple actually is or else at a "surrogate location" that is obviously related. Of course, if and when "this" refers to a part of my own body, "this toe" for example, what I would naturally do to create a physical disturbance at the relevant location is to activate the selfsame body part: "this toe" is "the very toe I am now wiggling."

Now it so happens that certain indexicals have an interesting property, namely that the activity of *speaking* them can by itself be the bodily activity that does the job of indicating what "this" is. When, for example, I say "now" (meaning "this time"), I indicate the time in question simply by making the sound at that very time. When I say "here" (meaning "this place where I am"), I indicate the place in question by moving my mouth in that very vicinity. And when I say "I" (meaning "this person"), I indicate the person in question by speaking with that very person's mouth. Indeed if I were to say "these lips," I would indicate the lips in question by moving those very lips. Thus these indexicals require no further act of indicating to make their meaning clear, since, when spoken, they issue from exactly the space-time location that is indicated.

But now, if I can indicate my lips to someone else by the self-sufficient act of saying "these lips" and moving my lips at the same time, what would I have to do to indicate my lips only to myself? In my own case I would not presumably have to say "these lips" out loud, for it would come to the same thing if I were to say it simply *sotto voce*. But, more than that, in my own case it would come to the same thing if I were simply to think the thought "these lips" and not say anything at all—provided of course that I still made a slight movement with my lips or at the very least I initiated some activity that points in their direction. And if this is true of lips, there is no reason why the same should not hold for any and every other part of my own body. Thus merely thinking "this toe" or "this hand" and making some slight movement with the relevant appendage would be sufficient to indicate the toe or the hand to myself—and for the thought to be self-indicating.

Or would it? We should be careful about this. For if I were merely to think "this toe" or "this hand," the thought as such would *not* be self-indicating in the way the indexical speech act is, unless the thought were to be in some way directly tied in with the movement of the body part in question in the way the speech act is. A thought that causally brings about the movement would do the trick, but a thought that just happens to be accompanied by an independently caused movement would not. In other words, a thought or indeed any other mental state will be self-indicating if and only if it both refers to a particular site in the body and produces a physical disturbance at the very site referred to. In fact for a thought, by itself, to indicate my toe it has to be a thought that reaches out and "thinks-my-toe-to-move."

What kinds of thoughts or other mental states are, or might be causally efficacious in this special way? It has been claimed—without very good evidence—that almost any act of "attending" to a body part will in fact reach out and automatically bring about at least a micromovement of the part in question: so that, if a person focuses his attention on his left foot, he makes at least a slight movement with the foot; on his tongue, he makes a slight movement with his tongue; on his right ear, he even makes a movement with his ear! (Try it: you will perhaps recognize that something of the sort does seem to happen.)

Yet of course it is not "attentional movements" but rather "intentional movements" that provide the best example: movements, that is, that form part of voluntary bodily activities wherein the executive self by an act of will commands a part of the body to do something. My foot may or may not move automatically when I focus my attention on it, but there can be no question that it moves automatically when I will it to move. Such bodily activities are therefore the paradigm examples of self-indicating states.

But now all we have to do to close the circle is to note that not only are these the paradigm examples, they are in the last analysis the only examples. For in fact any mental state that unites these two elements of referring to a site in the body and reaching out to create a disturbance at this site would belong to the class of bodily activities by definition—because this is precisely what a bodily activity amounts to.

Thus a state can be self-indicating (or, to return now to my original phrase, a state can be self-characterizing in respect of its location) if and only if it too is some kind of bodily activity. And since our starting point was that sensations also do it, we can conclude that sensations themselves are indeed a form of bodily activity. Only, now we have a firmer fix on what this really means: namely, that sensations themselves are reaching out to the site that they refer to and creating a physical disturbance at the relevant location.

Admittedly, as noted above, the "relevant location" could be a location on a map or plan—a surrogate location that is obviously connected to the true one—and therefore need not actually be the body part itself. If human beings possess an "inner model" of their bodies, the sensory activity that indicates the body could be a quasi-activity involving not the real body but this inner model. But the conclusion stands that, one way or another, sensations must be actively doing something to create a disturbance at "this body-related place, here and now."

In short, just as to wiggle my toe is to send an outgoing signal to my toe to wiggle (which is *why* and *how* the activity directly implicates my toe), so to feel a pain in my toe must be to send an outgoing signal to my toe to hurt (which is why and how the sensation does it too).

This has been a difficult argument to make, and possibly to follow. And, even if the thesis makes sense in the case of pain (and perhaps touch in general), there might still seem to be problems in extending it to other sensory modalities: to the case not just of feeling-a-pain-in-my-toe but, for example, of feeling-sweetness-at-my-nose or of feeling-redness-at-my-eye.

People do, as noted, actually say "my toe is hurting," or "my skin is itching," or "my face is burning," using language very similar to the activity-language of "my toe is wiggling." But they do not say "my nose is sweeting" or "my retina is redding." And indeed it has still to be asked what kind of centrally generated physical disturbance could be being produced at the nose or the eye.

But, having established the general thesis that sensations in the final analysis really *must* involve some sort of activation of the body surface, the way forward is clear. This thesis must be used to develop a story about the biological evolution of sensations.

20

PLUS ÇA CHANGE...

The beginnings of an evolutionary story were sketched in Chapter 3, where I suggested that the first function of sensations was—and remains—that of mediating an *affective response* to stimulation occurring *at* the body surface:

> [In the earliest animals] boundaries—and the physical structures that constituted them, membranes, skins—were crucial. They formed a frontier: the frontier at which the outside world impacted the animal, and across which exchanges of matter and energy and information could take place. Some of these stimulus events were, generally speaking, "a good thing" for the animal, others were neutral, others were bad. Any animal that had the means to sort out the good from the bad—approaching or letting in the good, avoiding or blocking the bad—would have been at a biological advantage. Natural selection was therefore likely to select for "sensitivity."
>
> Being "sensitive" need have meant, to begin with, nothing more complicated than being locally reactive: in other words, responding selectively at the place where the surface stimulus occurred. The first types of sensitivity would have involved, for example, local retraction or swelling or engulfing by the skin. Soon enough, however, more sophisticated types of sensitivity evolved. Instead of or as well as a stimulus inducing a local reaction, information from one part of the skin got relayed to other parts and caused reactions there, and, with different stimuli coming to elicit very different action patterns, the way was open for the animal's responses to become better adapted to its needs. Since information about the particular stimulus was being preserved and carried through into

the particular action pattern, the action pattern had come to represent the stimulus.

It was being suggested, then, that sensitivity evolved primarily as a means of *doing something about the stimulus at the point of stimulation*: at least to begin with, the animal both detected and responded to the stimulus with the same bit of its skin—the sensory epithelium was also the responsive epithelium, and the sense organ (if it deserved the name of organ) was also the effector organ. But while I went on in Chapter 3 to stress the subsequent *decoupling* of sensitivity and responsivity—which led eventually to two channels of representation, sensation and perception—my position on this issue has now changed. For we now have every reason to stress the *coupling that remains*.

The reason is the one elaborated over the last few chapters. Every sensation is still, even in contemporary human beings, felt as something happening "here" and "now" to "me." And this logically requires that the sensation (or the action plan that corresponds to it) continues to reach back to the point of stimulation—to indicate the "there" and "then" "to whom."

I suggest we should be thinking then, quite simply, in terms of an evolutionary continuum, as shown in Figure 5, in which, even as the sensory response became more complicated, some version of the original arrangement was still retained.

In the most primitive animals, the response to stimulation would have stayed entirely local: when, for example, the surface of an amoeba was touched, there would have been a spread of excitation directly across the cell membrane to produce a defensive wriggle by that part of the membrane. In a more highly developed animal such as an earthworm, the response would have come to involve signals traveling to and from a more centrally placed ganglion. And in human beings the response has come to involve signals going all the way from the body surface to the brain and back again.

Is there anatomical evidence to back up this scheme? I would say just that there is sufficient evidence not to discount it. All afferent sensory nerves in human beings do, as it happens, carry at least some efferent fibers, and even in the case of the eye some 10 percent of the fibers in the optic nerve conduct signals from the brain back to the retina (which means there are considerably more fibers going out to the retina than there are, for example, going out to the muscles of

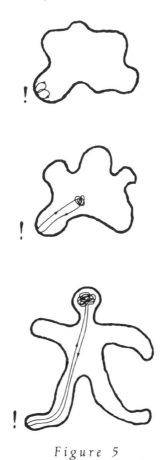

Figure 5

the hand). But I would say also that it would be a mistake to let the anatomical facts, as known at present, limit the discussion. There will be occasion later to tailor the theory to the physiological reality of human bodies.

The main proposal for the moment is simply this: that the activity of sensing, even in human beings, is a direct descendant of the primitive affective response. The "sensory loop" has gradually lengthened. Nevertheless an unbroken tradition links the sensations of modern human beings to those original amoeboid wriggles of acceptance or rejection. The more things have changed in the course of evolution, the more they have stayed much the same.

Biologists (and philosophers too) who want to understand contemporary facts do well to pay close attention to where things are descended from, their pedigrees.

Consider, by analogy, the remarkable case of the green turtles of the South Atlantic which swim two thousand miles to lay their eggs. It was not always so. One hundred million years ago, when only a narrow strip of sea separated the continents of South America and Africa, the turtles living off the South American coast laid their eggs just a short distance away on an island near Africa. Then continental drift got under way, and the African plate and the American plate started to move apart, opening up the vast Atlantic Ocean in between. What happened? The turtles' traditional feeding ground was located on the South American side, while their traditional breeding ground was on the African side. But rather than changing their ways, year by year they swam a little further eastward. The result was that today the turtles make an "unnecessary" journey, that—if we did not know its history—might seem biologically absurd.

I do not mean by this analogy to suggest that there is anything comparably absurd about sensations. But I do mean to suggest that if human sensations, following an age-old route, still reach back from the brain to the very point where the sensation is being felt, and if the activity that they perform there is descended from the affective responses of our far distant ancestors, we can expect *this* to be the key to what at a deeper level they still are today.

To take this further, however, we shall have to be much more specific—and, in particular, to address an obvious problem. If human sensations are descended from what were originally amoeboid wriggles of acceptance or rejection at the body surface, then how could there have come to be a sufficient variety of "sensory responses" to underlie all the richness of human sensory experience?

21

A LITTLE MIND MUSIC

Maybe *it is all* right for an amoeba, which probably does not enjoy a particularly rich sensory life: different sorts of "wriggles of acceptance or rejection" might indeed provide a sufficient basis for everything an amoeba can represent. But it is not—at least not obviously—all right for human beings, for whom there would seem to be vastly more ways of sensing a stimulus than there could conceivably be ways of "wriggling" in response to it.

At the end of the last century, certain scientifically minded psychologists attempted to assess the total number of sensations distinguishable by human beings. Edward Titchener counted 44,435 "elementary sensations," including 32,820 for vision, 11,600 for audition, and 1 (yes, only one) for sex.[116]

We need not accept these figures to appreciate that there would indeed be a considerable quantitative problem in mapping human sensations onto different kinds of bodily activity. But more serious is the qualitative problem. For what could be the crucial difference between performing a "red wriggle" at the retina, a "sweet wriggle" at the tongue, and a "ticklish wriggle" at the elbow? How could any outgoing signal from the brain to the periphery contain this sort of information?

My hypothesis may stand or fall on providing a realistic answer to these questions.

It may be helpful to introduce a change in terminology. Instead of talking about sensory responses, let alone about wriggles of acceptance or rejection, we ought to have a more specific word to name what I have been calling the "activity of sensing"—and preferably a word that also has the connotations of affect. Neologisms sound ugly, and no existing word is altogether right. Nonetheless I suggest that, even if it takes some getting used to, we call the centrally occurring activity "sentition," and the actual events at the body surface that flow from it "sentiments." Thus sentiments, in this usage, would be the name of the real physical disturbances that, *ex hypothesi*, occur at the place where sensations are being felt.

Let us suppose, then, that every distinguishable sensation in human beings does correspond to a physically different form of sentiment occurring at the body surface. Indeed let's suppose, for the sake of argument, that what it is for someone to *feel* a particular sensation is just for him to engage in the *appropriate form of sentition*—and to issue whatever instructions are required to create the relevant outgoing signal from the brain. And the question is: what features of these sentiments could correspond to the qualitative dimensions of sensory experience, and what features of the outgoing signal could encode them?

We have two pieces of evidence (perhaps just two) to go on. The first is the fact that in human beings there is, as we noted, an association between the "modality" of a sensation and the bodily location at which the sensation is felt to occur: so that people typically have visual sensations with the retina, olfactory sensations with the nasal mucosa, tactile sensations with the skin, and so on. The second is the fact that, even today in modern human beings, there is still at least a vestigial association between the "submodal quality" of a sensation and the way the stimulus is evaluated at an affective level: so that within the visual modality red light is typically exciting, blue light calming; within the tactile modality itches are irritating, tickles pleasurable; within the gustatory modality sweet tastes are appetitive, rotten tastes revolting; and so on.

Now, in relation to the first fact, note that each of the modality-specific areas of the human body looks very different under the microscope and does indeed have its own distinctive physical microstructure. Hence, when a particular area is implicated in sentition, it is likely that all the sentiments in this one area have a characteristic structurally determined form. So, it can be suggested that the *modality* of a sensation is directly linked to this *structural* dimension of the corresponding sensory response—with visual sensations being linked to the particular form of retinal sentiments, olfactory sensations to the form of nasal sentiments, tactile sensations to the form of skin sentiments, and so on.

In relation to the second fact, note that the way a person as a whole responds affectively to stimulation is likely to be correlated with the way that he responds (or at least his ancestors in the evolutionary past responded) affectively at his body surface. Hence sensory responses probably still retain at least the ghost of their original affective function, and different sentiments, occurring within the same area of the body, are likely each to have a characteristic functionally determined form, according to whether they are (or at least have been in the past) designed to welcome the stimulus, reject it, or whatever. So, it can be suggested that the *submodal quality* of a sensation is directly linked to this *functional* dimension of the corresponding sensory response: with sentiments that act to increase the stimulation having one submodal quality, those that act to decrease it another submodal quality, those that act to maintain it constant another, and so on over a wide range of more nuanced positive or negative affects.

This may not seem much to go on; but it is promising. If we think about larger-scale bodily activities, it is evidently just these two features—bodily location and function—that determine their "adverbial style." Thus the analogy noted earlier between feeling qualitatively distinct sensations and performing qualitatively distinct bodily activities continues to prove surprisingly apposite. The difference between feeling a tactile sensation at the elbow and a visual one at the eyes could be said to be a bit like the difference between performing a locomotory activity with the legs and an ingestive one with the mouth; and, within a modality, the differences between feeling sensations of pain, itch, and tickle could be said to be a bit like those between hopping, running, and skipping.

I shall not attempt to specify how this might actually work out in detail: partly because the suggestions I have just made are in certain respects (we shall see later in what respects) some way away from the final biological reality. But, as a purely abstract illustration, perhaps the wavy lines of Figure 6a can be taken to represent different sentiments occurring in different areas of the body surface, corresponding to sensations belonging to different sensory modalities; and those of Figure 6b can represent different sentiments within

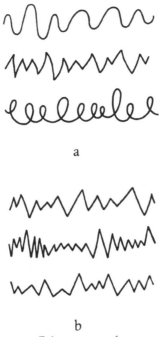

a

b

Figure 6

a single area that have different affective functions, corresponding to sensations with different submodal qualities.

I like this way of illustrating sentiments—as if they were literally waves of activity occurring at the body surface—because it suggests a musical analogy.

Think of a concert orchestra, spatially arranged with string instruments in one area of the stage, brass in another, woodwind in another, percussion in another, and so on. And imagine that this orchestra has a conductor—a real maestro—who not only sets the pace and brings in particular instruments on cue but actually gives each individual player instructions as to what actions to perform.

Suppose that the orchestra corresponds to the surface of someone's body, with each section being a different sensory area, and that the conductor corresponds to the source in the brain of outgoing sensory signals. Suppose further that the playing of a particular combination of notes on a particular instrument in this ensemble corresponds to a particular sensation, and that the conductor's role in creating this instrumental activity is equivalent to the brain's role in creating the sensation.

Then the modality of the sensation would correspond to the style of playing that is required by the structure of the instrument: in other words, the manner in which an instrument in this section of the orchestra has to be handled—fingered, bowed, blown, plucked, etc. And the submodal quality of the sensation would correspond to the actual combination of notes that the playing is intended to produce.

So that, for example, the tactile modality might correspond to the woodwind style, the visual modality to the strings style, the gustatory modality to the percussion style and the auditory section to the brass style. And within the tactile modality, itch might be a C minor chord on a flute, warmth an E flat chord on a bassoon, tickle a C major on an oboe.

Figure 7 represents this theory of sensations. Note the inner conductor, ''I.''

Where does the conductor get his own program for the activities he is directing? Well (unless he is dreaming), he gets it from information that he receives *from* the sense organs. However, this information does not of itself result in the making of music—any more than does a musical score. It is what the conductor does with it that matters.

Figure 7

· 22 ·

SPECIFIC NERVE
ENERGIES?

Some of this may perhaps be starting to sound topsy-tur-vy—especially what was said at the end of the last chapter about the incoming information from the sense organs having no "musical richness" in itself. When developed further, the hypothesis will however prove to have considerable virtues (once we have dealt with some potential flaws). But before going on to find out what can be done with it, it now needs to be put in the context of more traditional ideas.

The standard theory about sensations is, I suppose, the precise opposite of the one I have described, since it does put all the weight on the nature of the input to the brain rather than any output from it. In particular it assumes that the modality of a sensation is determined in the first place by the anatomical arrangement of the *incoming nerves:* so that, for example, if a signal comes in via the optic nerve and excites the visual cortex this is enough to ensure that the sensation is a visual one. In terms of the musical analogy, it would be as if there is someone inside the head listening to the music rather than producing it, an inner receiver rather than an inner conductor, who when he receives a message at the part of his brain supplied by the optic nerve experiences it as the sound of "visual strings," while when he receives it at the part supplied by the auditory nerve experiences it as the sound of "auditory trumpets."

This so-called doctrine of specific nerve energies was put forward by Johannes Müller as long ago as 1834. Here is a recent summary of the idea by an Oxford professor writing in the *Encyclopedic Dictionary of Psychology:* "Sensory quality depends on which nerve is stimulated. . . . Any kind of activation of the auditory nerves will produce auditory sensations, because the nerve goes to the auditory system of the brain. Similarly activation of the optic nerve produces visual sensations because the optic nerve transmits information to the visual system of the brain."[117]

The facts here are of course correct—if the auditory nerve is stimulated with electric current, for example, the subject may have a sensation of ringing in his ears but will never have a visual sensation, whereas if the optic nerve is stimulated by the same current he may experience flashes of light but will never have an auditory sensation. But I said above that I only "suppose" this idea is the opposite theory to mine because actually I do not think it should count as a theory at all. It certainly provides no sort of explanation for *how* sensations come to have the quality they do.

"Activation of the auditory nerve produces auditory sensations (rather than visual ones) because the nerve goes to the auditory system of the brain"! Someone might as well claim that feeding corn to chickens produces clucking sounds (rather than mooing sounds) because the corn goes to the "chicken system" of the farmyard, or that dialing 911 produces a policeman at the door (rather than a Chinese takeout) because 911 calls go to the "police system" of the telephone exchange. Even if correct, the explanation would be vacuous, so long as the workings of the "system" remain unexplained.

An explanatory theory of sensations cannot just take it for granted that different systems each, as it were, do their own systematic thing with the input they receive—when it is precisely this systematic thing that needs explaining. Rather, it must address itself to the nature of what each modality-specific system *goes on to do next*. Ideally the theory ought to provide good reasons for why the "auditory system" goes on to produce sensations that have just the auditory quality they do, whereas the "visual system" goes on to produce sensations that have the visual quality they do, and so for the other modalities. But if it cannot do that, it ought at the very least to offer suggestions about how what the auditory system does differs in a relevant respect from what the visual system does.

The fact is, however, that neither the doctrine of specific nerve energies nor any modern variant of it has anything to offer on this score. The recent literature in cognitive science or neurophysiology hardly even addresses the question of what produces the qualitative difference between different sensory modalities. If one were to ask most contemporary scientists to hazard a guess, they might perhaps mumble something about the "processing of information" being done in a modality-specific manner. But when pressed they would probably admit that they cannot even imagine how different kinds of information processing could do the job. There are only so many ways of passing impulses backward and forward among nerve cells, and none of them would seem adequate to underlie the difference in experience between seeing red and feeling pain. Remember the glum warning from Colin McGinn, quoted at the beginning of the book: "You can't get the 'qualitative content' of conscious experience— seeing red, feeling a pain, etc.—out of computations in the nervous system."

If, however, the standard theory has nothing much to offer here, can my hypothesis actually do any better? I would say that by focusing not on what goes *in* to the sensory systems but on what comes *out*, it is certainly in with a chance.

For a start the hypothesis suggests that the ways in which sensations differ must in the final analysis be ways in which the corresponding sentiments can differ. It thus shifts the problem away from information processing as such to a more restricted but more promising domain. It is more promising because we already have a model for how bodily activities on a larger scale can be almost as far apart in their "adverbial quality" as are sensory modalities. Perhaps not everyone would agree that blowing down a trumpet with one's mouth and playing a fiddle with one's hands are in such different leagues. But, for a grosser analogy, consider the differences between eating, dancing, speaking, and digging the garden: while it is easy to imagine a string of intermediate activities within each category, such as from dancing the tango to dancing the mazurka or from eating figs to eating turkey, there is arguably an absolute disjunction between dancing the tango and eating figs.

Moreover, this hypothesis opens up the possibility of arriving closer to what I called an "ideal" explanatory theory of sensations, namely an explanation that gives good reasons why the output from a sensory system should have just the quality it does have. For I think it may be possible, in principle, to establish logically necessary correspondences between the form of particular sentiments and the quality of particular sensations—based on formal *resemblances* between them.

I am not saying that anything I have yet suggested comes near to doing it. For I can, I admit, think of no *a priori* reason why for example a sentiment that has a retinally determined form should resemble a visual sensation, while a sentiment that has an aurally determined form should be auditory; nor why a retinal sentiment that is affectively alarming should resemble a red sensation, while one that is pacifying should be green. Nonetheless if there *is* a relationship between the form of sentiments and the quality of the corresponding sensations, then—unless God is playing dice with mind–body relationships—we can take it that the relationship *must be* nonarbitrary. It has to be a "motivated" relation, as the semiologists would say. And when we have a decent theory of sensations, it will come to be seen as motivated and nonarbitrary.

If and when we have this theory we shall be approaching what many theorists have thought impossible: an "objective phenomenology" that links sensory experience directly to what is happening in the brain and body. We should in principle be able to *deduce* what a person is experiencing from observations on his brain and body. And if we can do this for another human being, we should be able to do it also for a bat . . . or a wombat . . . or for that matter a robot. We might even come to see how a philosophically minded robot could deduce the same for us.

We may not, in fact, be nearly there yet. But we have stolen a march on other theorists by even anticipating that there is a "there" to get to.

When Howard Carter, excavating in the Valley of the Kings, broke through to the tomb of Tutankhamen, and peered through the peephole he had made, his companions asked him, "What do you see?" He answered, "Wonderful things." But, then, he had to stand back and continue the heavy work of knocking down the wall.

· 23 ·

SMOKE
WITHOUT FIRE

William Blake, the poet, would not much have liked the line of reasoning so far. "Mental things alone are real," he wrote, "I question not my corporeal or vegetative eye any more than I would question a window concerning a sight. I look through it and not with it."[118] Or, as he objected again, in a later poem:

> This life's five windows of the soul
> Distorts the Heavens from pole to pole,
> And leads you to believe a lie
> When you see with, not through, the eye.[119]

A lie? There have not, I think, been any lies involved in the argument that I have been putting forward. Even so, the point is surely coming when it may be necessary to take account of certain awkward truths.

Do I really want to claim that sensations are felt *with* the body surface: that pain sentiments have to occur *at* the skin, gustatory sentiments *at* the tongue, and indeed visual sentiments *at* the eye?

Perhaps I might *want* to claim it for all the reasons given earlier. But the tragedy of science, it has been said, is the slaying of a beautiful hypothesis by an ugly fact. And I shall not of course insist on this Mark-1 version of the theory if it is evidently wrong.

The ugly fact (and maybe not the only one) that has been waiting in ambush for the hypothesis as stated is the fact that in some circumstances people can have sensations in parts of their bodies that do not physically exist.

The most telling—because most dramatic and horrible—example is that of "phantom limbs." Phantom limbs are imaginary limbs that persist after a real limb has been amputated. Immediately following an amputation, and often for months or even years afterward, the patient may report that he has a definite sense that the limb is still a part of him. As Ronald Melzack, an authority, describes it: "The phantom limb is usually described as having a tingling feeling and a definite shape that resembles the real limb before amputation. It is reported to move through space in much the same way as the normal limb would move when the person walks, sits down, or stretches out on a bed. . . . Although tingling is the dominant sensation, amputees also report a variety of other sensations, such as pins-and-needles, warmth or coldness, heaviness, and many kinds of pain. About 35 per cent of amputees report pain at some time. Fortunately, the pain tends to subside and eventually disappear in most of them. In about 5–10 per cent, however, the pain is severe and may become worse over the years. It may be occasional or continuous, and is described as cramping, shooting, burning or crushing. . . . The pain is felt in definite parts of the phantom limb. A common complaint, for example, is that the phantom hand is clenched, fingers bent over the thumb and digging into the palm, so that the whole hand is tired and painful."[120] The pain goes on occurring despite the fact that the original wound has healed completely and incoming pain nerves are no longer active.

Now, it must be clear that if my initial hypothesis were right such phantom sensations should simply not be possible. Phantom pain obviously cannot be felt *with* the amputated limb. A nonexistent foot should not be able to hurt (note the active verb) any more than it can wriggle: no foot, no possibility of pain sentiments occurring in the foot, hence no pain sensations. Yet try telling that to the subject of the pain! A sixteenth-century surgeon, Ambroise Paré, remarked: "Verily it is a thing wondrous strange and prodigious, and which will scarce be credited, unless by such as have seen with their eyes, and heard with their ears, the patients who have many months after the cutting away of the leg, grievously complained that they yet felt exceeding great pain of that leg so cut off."[121] Third-person, theoretical skepticism obviously has to retreat in the face of undeniable first-person suffering.

Phantom sensations may also occur after loss of the eyes. Although, so far as is known, there is no visual equivalent of phantom limbs as such—a fully formed phantom visual field following destruction of the eyes—nevertheless the sudden loss of both eyes does not bring an end to visual sensations altogether. Although the cases are fortunately rare, and have not been systematically studied, there are reports that for a short time afterward the victim may experience a variety of sensations in his visual field, such as sparks of light, shooting stars, flames, or clouds. More common are cases where the eyes, while still intact, have been cut off from the brain by damage to the optic nerve. And in these cases more complex illusions are reported. For example, in an eighteen-year-old woman who became completely blind as the result of an operation to remove a tumor impinging on the optic nerve: "Following discharge from hospital she began to see 'light'; later she saw moving objects like snakes and also colours, then scenes appeared composed of persons and objects; these annoyed her, prevented her from sleeping, and interfered with her daily activities."[122]

As in the case of pain, therefore, there is clinical evidence that the experience of visual sensations cannot depend on sentiments that actually take place at the retina. We might however have drawn the same conclusion without going so far afield. For, if all we want is evidence that people can have sensations in a small part of the visual field that does not exist at the eye, we have only to consider our own retinal "blind spots."

There is a naturally occurring hole in each of the retinas of the two eyes, about a millimeter square, in the region where the optic nerve leaves the eye. Since light falling on this hole goes undetected, any part of the retinal image that falls there vanishes from sight.

X   BLIND
                                                           SPOT

The consequences are easily demonstrated. Close your left eye and look with the right eye at the X, with the page about 12 inches away. If you move the page backward and forward a little, you will find there is a position where the words BLIND SPOT disappear. (If you now open your left eye the words will appear again: the blind spots of the two retinas do not overlap.) The point to note is that the blind spot is not experienced as being an empty area. When the words disappear the white background spreads in to fill the gap; and if the page were red or blue or green, the filling in would be of the corresponding color.

The point is, again, that such phantom sensations in the blind spot cannot be felt *with* the eye. They ought therefore, according to the Mark-1 theory, not to happen: no retina, no visual sentiments at the retina, no sensations of light.

There is evidently no way out of this except to modify the theory. If the Mark-1 theory cannot cope, we need a Mark-2 theory that, while retaining the essential features of the previous version, is better adapted to the facts.

The two features that have to be retained are these. First, the idea that there has been an evolutionary continuum in the development of sensory activity from amoebae through to human beings. And, second, the logical requirement that in order for sensations to be self-characterizing in respect of their location they must still be reaching back to create a physical disturbance at the site where they are felt.

However, if the site they reach back to is, in human beings, no longer necessarily the actual body surface, where is it?

Recall that in the earlier discussion of the logical status of indexicals I was prudent enough to insert a kind of escape clause: "When 'this' refers to a part of my own body, 'this toe' for example, what I would naturally do to create a physical disturbance at the relevant location is to activate the selfsame body part: 'this toe' is 'the very toe I am now wiggling.' . . . [But] the 'relevant location' could be a location on a map or plan—a surrogate location that is obviously connected to the true one—and therefore need not actually be the body part itself. If human beings possess an 'inner model' of their bodies, the sensory activity that indicates the body could be a quasi-activity involving not the real body but this inner model."

The escape was this idea of an "inner model of the body"—a model in the brain. But what exactly could such an inner model be?

Presumably, if the model is to be the basis for the physical disturbances that underlie acts of indication, it has to be more than a purely "abstract" or "conceptual" model. Indeed presumably the model must be some kind of physical structure in its own right: so that for every location on the real body surface where sensations are felt there is in fact a physical location on the model body where the corresponding sentiments can occur. What is more, this surrogate location must be "obviously connected" (as I put it above) to the real one.

But what precisely could that mean? By virtue of what could a location in the brain be "obviously connected" to a location on the body surface?

There is, I think, no choice but to go for the strong interpretation here: it must mean that when something happens at this surrogate location in the brain it will seem to the subject as if it were happening at the corresponding location on his body surface: a physical disturbance at the model toe will have to be subjectively indistinguishable from a disturbance at the real toe.

But how could this be brought about?

The obvious answer would be that the surrogate location is itself located on the path of—or more probably at the terminus of—an *incoming sensory nerve* coming from the relevant part of the body surface. In other words, the surrogate location of, for example, my left big toe would be the point at which the incoming sensory nerve from the toe arrives at the "toe area" of the tactile cortex of the brain; and in general the surrogate locations for all other parts of the body surface would be the corresponding points of arrival at the cortex of nerves from the skin, mouth, eyes, ears, etc.—with, in particular, the visual cortex representing the retina, the auditory cortex the basilar membrane, and so on.

If this is right, the inner model of the body would be simply this input-defined cortical map. And where I wrote above of "the activity that indicates the body" being "a quasi-activity involving not the real body but this inner model," we could now assume that the quasi-activity reaches out to and has its effects at the sensory cortex as such.

I say this is the obvious answer. It is indeed a simple answer. But none the worse for that. For I suspect it is the only (non-tendentious) answer that will do the trick: the requirement being that an act of indication at body point P should be replaceable in principle by one at brain point p.

It would make sense then to suggest the following revision for the Mark-2 theory.

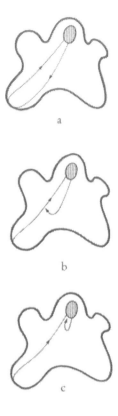

Figure 8

Sensory information arrives at the brain via incoming sensory nerves and, as before, the subject responds by directing a sensory response back out toward the body surface. But I now propose that, in the course of evolution, the target for these sensory responses has progressively shifted inward from the real body surface along the path of the incoming sensory nerves. So that there has been, as it were, a short-circuiting of the sensory response, a closing of what I earlier called "the sensory loop." Where once the response reached all the way back to the point of stimulation (Figure 8a), today it ends at the surface of the brain (Figure 8c).

How would this new version of the theory cope with the paradoxical examples earlier in this chapter? Evidently the preconditions for having a sensation would have significantly altered. Sensations—even illusory sensations—instead of depending on the existence of the real body surface would now have come to depend on the existence of the cortical sensory projection areas.

This being so, there would no longer be any great theoretical problem about phantom sensations occurring after amputation of a limb or loss of the eyes, since the sensory cortex that once received input from the missing body part would still be intact and hence the surrogate location for painful sentiments or visual sentiments would still exist. It is true that phantom sensations in the blind spot might still seem something of an anomaly, since they would have to depend on there being a cortical area corresponding to a retinal area that has never existed at all. But there is in fact a natural explanation here, namely that the two eyes send overlapping projections to the cortex and their blind spots occur in different places, so that each of the separate blind spots is "covered" at the visual cortex by a location that receives its input from the other eye.

One should of course expect that loss of sensory cortex as such would lead to the complete loss of both normal *and* phantom sensations. And so in fact it does. After destruction of the visual cortex, for example, patients not only lack all normal visual sensations, but (unlike the young woman with optic nerve damage I mentioned earlier) they do *not* experience spontaneous visual phantoms, nor do they have visual imagery, and nor—when the destruction is complete—do they have visual dreams. They may still have the rudimentary capacity for blindsight: but this, as we saw, is basically a perceptual capacity not a sensory one.

The revised version of the theory is therefore able to handle the potentially awkward clinical evidence with relative ease. (Luckily—but is it luck?—it also ties in with the evidence about sensory imagery involving the cortical projection areas referred to earlier on.)

The original theory of sensations as bodily activities has undergone a fairly radical revision—to such an extent that it may seem no longer to count as the same theory.

I am still maintaining that to have a sensation involves the making of a "sensory response." But this response, which began its theoretical life as a real *bodily* activity, has now become some sort of *brain* activity. As William Blake might have put it (if he had been following the discussion): "corporeal sentiments" have become "cerebral sentiments."

Figure 9 shows more explicitly what the new theory amounts to. While the original version proposed the arrangement shown in (a), the revised version proposes that shown in (b). Where once the inner conductor had a full bodily orchestra to play with, now he has only the sensory cortex at his disposal.

I am suggesting that this theoretical revision corresponds to an evolutionary revision. The cerebral sentiments of Figure 9b are direct descendants of the corporeal sentiments of 9a; and many of the original considerations will still apply. Yet the whole point about evolution is that, however great the biological continuities, things really do change. In fact, despite everything said earlier about the importance of pedigrees, it is surely conceivable that step-by-step evolutionary progress might have resulted in a complete turnaround in function or in meaning.

Great store has been placed up to this point on the argument that sensations must actually be *doing* something at the place that they are felt: that sentiments really are—or were—a form of *action* occur-

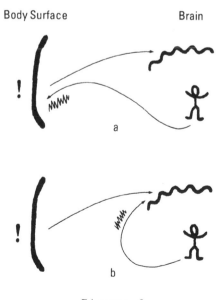

Body Surface   Brain

Figure 9

ring at the body surface. It may, however, be difficult to sustain this emphasis much longer. Cerebral sentiments, even though they may be descended from the original amoeboid wriggles of acceptance or rejection, are clearly no longer any form of wriggle in themselves. In fact it looks as though, instead of involving any sort of action, they have become merely patterns of nerve impulses terminating at the surface of the cortex.

Terminating and *doing* what? While an organism can wriggle its skin, it is by no means obvious how it could wriggle its sensory cortex. And, even if it could do so, it is quite unclear what it would achieve.

We undoubtedly have an interesting new puzzle here. But in point of fact we also have new clues. While it may be unclear what "wriggling the brain" would achieve according to the theory as it stands, it is perfectly clear what it would have to achieve if the theory is to contribute to solving the mind–body problem. For, in moving on theoretically from bodily sentiments to cerebral ones, we have moved on in evolution from archaic organisms like amoebae to conscious creatures like ourselves. And in our own case—even if we cannot speak for an amoeba—we know that one of the results of sensory activities is that we end up *feeling* a sensation: that is, we end up having the *conscious* experience of a pain in our toe, a scent in our nostrils, or whatever.

We know, in other words, what the theory of cerebral sentiments has to deliver. And all that is now needed is the means.

TIME PRESENT

I *proposed* "for the sake of argument" in Chapter 21 that "every distinguishable sensation in human beings does correspond to a physically different form of sentiment" and that "what it is for someone to *feel* a particular sensation is just for him to engage in the *appropriate form of sentition*—and to issue whatever instructions are required to create the outgoing signal from the brain."

However, this proposal was perhaps a somewhat rash one. If the subjective experience of having a sensation were to consist "just" in issuing instructions from a central site, then, if this "just" means what it ought to mean, it would seem to imply that *all* that matters is "instructions"—and the sentiments *as such* drop out of consideration. In which case, so far as subjective experience is concerned, much of the preceding discussion would have been beside the point.

I can conceive of someone arguing like this:

"Let's grant, as you want, that sensations involve a sensory response, with a signal being sent from a central site back out to a peripheral location (originally to the body surface itself but later to a surrogate location at the cortex of the brain). Nevertheless, once the signal has left the central site its mental work is done; and what happens to the signal after that obviously cannot influence the experience of it.

"The point—and I know that you'll appreciate this—is a logical one. What becomes of something in the *future* cannot change its *present* meaning. If, for example, you write a letter, address it to a particular house, and put it in the mailbox, the act of sending the letter is complete: and whatever becomes of it later can have no bearing on the meaning of the original act. Even if the letter were lost, the *intention* to send the letter would have been there.

"The same point could be made with a computer. When you set up a computer to display a circle on the screen, the computer sends an outgoing signal that produces the equivalent of 'circular sentiments' at the screen. If you now turn the screen off, but leave the computer running, the circle disappears. But the central processing unit of the computer is still issuing the relevant 'instructions' and sending them down the relevant wires. So the computer still 'thinks' it's drawing a circle.

"Now, take your inner conductor. Like the computer's central processor, this conductor presumably knows nothing about what happens to his instructions *after* he issues them. So sentition can take place independently of any actual sentiments occurring. And it follows that much of the discussion you have been having in the last few chapters about where sentiments take place and what they do there and how they correspond to particular sensations is a red herring.

"I'm not saying that sentiments don't actually exist. I agree with you that the instructions for them have to exist, and the instructions for one sentiment have to be different from those for another. And of course the instructions have to be directed somewhere. But the point is that what they do when they arrive there will be of no consequence so far as inner experience is concerned.

"What I'm claiming, if you like, is that 'unfulfilled sensory activities' can play just the same mental role as real ones. All that matters is *intention*. And I say 'if you like' because there have been strong hints that this is actually your own opinion—not only in that passage about 'just issuing instructions' but also even earlier on. Indeed the idea of intentional activity—of 'action unfulfilled'—was there right back in Chapter 7, when you gave that quotation from Coleridge about 'visual appetite': 'Sometimes when I earnestly look at a beautiful object or landscape, it seems as if I were on the brink of a fruition still denied . . . even as a man would feel who . . . leaps and yet moves not from his place.' "

· · ·

Touché. There is undeniably something right with this (even though I would say it was a bit unfair to bring back Coleridge). But—fortunately—there is something major wrong with it as well.

What is right and what is wrong?

It is the very concept of "instructions" that is the joker here and that is getting both me and this sparring partner into difficulties. What precisely does the concept mean? What makes an instruction count as an "instruction"?

In general it must surely be correct to link the concept of instructions to *intention*. Nothing can count as an instruction unless it is an instruction *for* something or *about* something. Instructions are essentially *forward looking;* they have to have an *anticipated* outcome. No signal, no matter what its *effects* are, can be an instruction unless its sender *already has these effects in mind*.

Imagine, for example, the following string of numbers being transmitted as a signal down a wire: 0462742065. Since the number happens to be my home telephone number, then, if the signal were sent from a phonebooth to the telephone exchange, the *effect* would be to create a ringing sound on the receiver now sitting on my desk. But this does not mean of course that the signal would necessarily constitute an *instruction* to that effect: the instruction to "Call Nick." Indeed it would only count as this instruction if the sender did have the specific *intention* of "calling Nick" in mind. If, to the contrary, the sender were merely dialing at random, and did not know what he was doing, then, even if the same signal went down the same wire and had exactly the same causal effects, it would not constitute this instruction nor necessarily an instruction of any kind at all.

Now, granted that this is the general rule—that a signal, just on its own, *cannot* amount to an instruction—the same presumably must apply to the signals that result in sentiments. A pattern of nerve impulses traveling either to the body surface or the cortex cannot just on its own constitute an instruction for a sentiment, since there can be nothing anticipatory or intentional about such a pattern of impulses *per se*.

But in this case, if we still want to suggest—as I did originally—that sentition consists *just* in "issuing instructions," we are apparently in a somewhat embarrassing position. For who or what are we going to make responsible for the intentionality?

Are we to suppose it is "I," the "inner conductor," who plays the necessary forward-looking role—anticipating what sentiments his signals are intended to create?

The answer has to be that this won't do. Or at least that it will not do as things stand. For, as things stand, the last thing we ought to be supposing—if we value theoretical respectability—is that the inner conductor is capable of anticipating or intending anything. The inner conductor is, after all, a mere functionary. His role in the theory is not himself to have a mental life but to help us explain mental life—not to be conscious but to explain consciousness. If we once start crediting this inner conductor with his own intentional states we shall be heading for an infinite regress.

All sorts of problems are looming now—of the kind that excite analytical philosophers. But rather than being drawn into a discussion on their terms, we must break out with a new line of our own.

What was right about the argument above was the assumption that instructions are intrinsically forward looking. What was wrong, I think, was the deceptively straightforward argument that followed on: that *because* they are forward looking *their actual outcome does not matter*. It may be that precisely the opposite is true.

To return to the example of the stranger dialing the number of my house: we assumed that he did not know what he was doing and so was in no position to anticipate the effects of the signal he transmitted down the wire. We might however look at it another way. The fact that he did not immediately know what he was doing would have been no bar to his getting to know *later*. Indeed we can take it that he would have *got to know* what he *had* done just as soon as someone answered the telephone and said "Nick Humphrey here."

Then, could it be that the return message would have rapidly transformed the meaning of his original signal? Could this signal have become in retrospect the instruction to call Nick? Could it have become in retrospect the instruction in prospect to call Nick? And, if so, would we have a model for how, in general, nonanticipatory signals could qualify as "instructions" by virtue of the return messages they might set up?

This sounds odd. It would seem to require some kind of backward causation. And such backward causation is exactly what our adversary was objecting to earlier: "What becomes of something in the *future*," he said, "cannot change its *present* meaning." And this, he insisted, was a logical point.

Logical it may have been . . . but, then again, maybe it was not entirely logical. For it could be argued that it all depends on what is meant by "present meaning": in particular, on when the "present" happens and on how long the "present" lasts.

Suppose the present were to be stretched out a bit. Suppose it were to last long enough for the present and the past to overlap. Suppose that, in T. S. Eliot's words:

> Time present and time past
> Were both perhaps present in time future,
> And time future contained in time past.[123]

Suppose indeed that human beings travel through life as in a "time ship," that like a spaceship has a prow and a stern and *room inside* for us to move around.

Well, in that case we would not be talking about the "present" as a physicist defines it. We might however be talking about the "subjective present" as we actually experience it. The "physical present," strictly speaking, is a mathematical abstraction of infinitely short duration, and nothing happens in it. By contrast the "subjective present" is arguably the carrier and *container* of our conscious life, and everything that ever happens to us happens *in it*. (It is clear that Daniel Dennett and Marcel Kinsbourne, in a recent paper, are also thinking on these lines.[124])

Consider the diagram below. The Roman numbers represent physical time, the Arabic numbers represent subjective time. The "physical present" lasts no time at all, so that when for example physical time VI arrives, physical time V has passed. By contrast the "subjective present" lasts, let's say, three units, so that subjective time *5* persists right through to subjective time *7*.

```
....III.... ....IV.... ....V.... ....VI.... ...VII... ...VIII...
                  − − − →
             Physical time


   1 2    2 3    3 4    4 5     5 6     6 7
....3.... ....4.... ....5.... ....6.... ....7.... ....8....
                  − − − →
            Subjective time
```

Then—to return to our problem about sentiments—if the signal for a sentiment (or telephone call) were to go out at time V and a return message were to come back at time VI, the outgoing signal and the return message would both be a part of the same subjective present between times 6 and 7. And if they were contemporaneous in this way, there would be nothing illogical about the latter affecting the present meaning of the former.

In which case we might now be permitted to suggest that to have a sensation is not, after all, *just* to issue an instruction, but rather "to issue a potential instruction and receive a confirmatory answering signal within the scope of the subjective present." The intentionality would have been established neither in retrospect nor in prospect but "in transpect": for the anticipated outcome and the actual outcome would be rolled in one.

But before this gets too elevated I should bring it to earth with what is actually a perfectly mundane hypothesis.

I asked earlier: what do cerebral sentiments *do* (assuming they do anything at all)? In the light of this discussion, a new answer becomes apparent, an answer that was already semiapparent in Figure 9 of the last chapter. It is that what the cerebral sentiments of Figure 9b do is, as it were, to tickle the incoming sensory nerves. They thereby set up a recurrent *feedback loop*—with the result that the outgoing signal and the return message meld into a larger, longer-lasting process.

There is nothing mysterious about a "feedback loop." "Feedback" occurs when the output from a system influences the input to the system; and a "feedback loop" comes into being when, in addition,

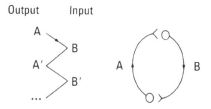

Figure 10

the input influences the output and a circle of causation is established.

Figure 10 shows a loop like this. Output A gives rise to input B, input B gives rise to output A′, output A′ to input B′, input B′ to output A″, and so on.

Since the activity in such a loop is self-propagating, this Ping-Pong exchange between input and output might in principle go on indefinitely. But in practice the process is likely to be damped. In particular, in the case of information flowing around a loop, some of the information is almost certain to get lost in the course of each circuit and the noise level to increase.

The rate at which the circulating signal decays will depend on the overall "fidelity" of the circuit. And two main factors are likely to affect this. First, how much of the information in the output actually gets returned as information in the input, and vice versa. Second, how much information gets lost along the outgoing and incoming pathways. In general, the tighter the coupling at each end—from output to input and from input to output—and the shorter and less noise prone the pathways, the longer the life of the signal going around the loop will be.

The possibility of sensory responses creating this kind of feedback loop was of course there from the beginning. In fact, it was not just a possibility but a fair certainty: for feedback is what responding *affectively* is all about. To "like" a stimulus is to respond to it in such a way as to keep up or increase the stimulation, and to "dislike" it is to respond in such a way as to keep down or reduce it. When a primitive amoeba for example responded with one of those wriggles of acceptance or rejection in the region of the stimulus, the effect—

Body Surface   Brain

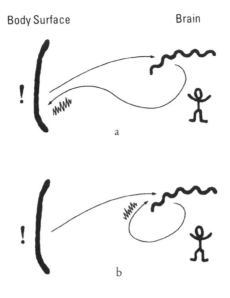

Figure 11

indeed the goal—of this response was precisely to influence the stimulus conditions to which it was responding. The makings of a feedback loop were therefore well in place.

We have to consider, however, how rapidly the activity in the sensory feedback loop would be likely to decay. And for this it may be helpful to bring back the diagram from the end of the last chapter, but this time with the completed loops sketched in.

In the early days, as represented by Figure 11a, we can take it that the loop would have had very low fidelity. One reason is that the loop was relatively long and probably relatively noisy. But another and much more significant reason is that the sensory responses were genuinely bodily activities, and the loop had to be completed via the outside world. The organism had to do something externally to change the input: it had, for example, to swim away from the source of stimulation, or suck it, spit it out, embrace it, kick it, or whatever.

In these circumstances the coupling between output and input could only have been relatively crude, and very little detailed information about the form of the response would have been transferred back to the sense organs. While the amoeba's wriggle, for example, would certainly have modified the input, the precise shape or dynamics of the wriggle would not have been preserved in the return message. Hence there would have been no real possibility of the

information about sentiments going around and around the loop—
and thus no possibility yet of the sensory activity being, as it were,
kept alive for an extended time by feedback.

However, with the evolution from corporeal to cerebral sentiments the situation was transformed. As sensory responses at the
body surface gradually got replaced by responses targeted on
the incoming nerves and eventually on the sensory projection cortex,
the result would have been not only a shorter loop but also a much
tighter coupling of output to input.

Admittedly there is no telling what the effects would have been
when, to begin with, the sensory response simply "tickled" the
incoming nerve. But in the course of evolution this tickling would,
we may assume, have become more and more specifically communi-
cative. The result would have been that eventually much of the
detailed information about the signal sent to produce the sentiment
at the sensory cortex would be being preserved in the signal returning *from* the cortex. And hence the signal in this purely "cerebral
sensory loop" could now reverberate for a considerable time before
it died away.

Supposing, then, that such reverberating feedback loops actually
exist in our own brains, we can return to the problem of "instructions" and the "intentionality" of sensory activity.

The problem arose with the insufficiently thought out suggestion
that to feel a sensation is "just to issue an instruction for a sentiment": for it was not obvious how the signals that give rise to
sentiments could ever count as an instruction *for* anything—unless,
that is, there were to be some sort of "backward causation."

But consider again the generic feedback loop of Figure 10. When
we have A causing B causing A' etc., we do *not* of course have
backward causation by B of A. But what we do have is forward
causation by B of A'. So, while it is true to say that the As as a whole
are the causes of the Bs as a whole, it would be equally true to say
that *over a long-lasting sequence* the Bs as a whole are the causes of
the As as a whole.

Thus what we have, oddly enough, is an overall blending of cause
and effect: the As which are the causes of the Bs are also the effects
of the Bs. And if, now, we identify this overall long-lasting sequence
with the "extended present," we have a situation where the As at the

time they go out are already—in the present—under the influence of the Bs they are going to give rise to. Hence the As, from being merely the signals that cause the Bs, have indeed become signals *for and about* the Bs.

I ought however to be more specific. Suppose A, A′, A″, etc. are the signals issued by the inner conductor that create red sentiments at the visual cortex, and B, B′, B″, etc. are the return signals to the inner conductor that the red sentiments are in fact occurring. And suppose, for the sake of argument, that the fidelity of the loop is such that the life of the activity set up by a flash of red light at the retina is about a tenth of a second: in other words that the circulating signal lasts about a tenth of a second before being lost as noise.

Now, if this tenth of a second corresponds to the subjective present, this would mean that throughout this present the inner conductor would be in the business both of issuing repeated signals for the red sentiments and receiving repeated confirmation of what the signals are doing. According to the analysis just given, the outgoing signals would thereby be transformed into signals *for* red sentiments. And, going by the updated criterion I suggested earlier—that "to feel a sensation is to issue a potential instruction and receive a confirmatory answering signal within the scope of the subjective present"—the subject would then be *feeling* the sensation of red light.

It would be nice to put some phenomenological flesh on these bare bones.

I assumed, to make the example relatively simple, that the incoming signal was brief—a flash of light. If the incoming signal were to be longer lasting, the situation would certainly be much more complicated because of the overlap that would be likely to occur between the current and recurrent inputs. Nevertheless we can safely assume that, when the stimulus persists, the sensory activity instead of dying away continues to reverberate and usually reaches some kind of equilibrium. We should expect therefore that, with a longer-lasting stimulus, the subjective sensation too would usually stabilize.

The possibility exists however that, if there were to be summation in the loop, the activity might *not* reach an equilibrium. We might expect there to be circumstances where the activity would rally to a crescendo, or hunt up and down in an oscillatory fashion. I can think of no obvious examples of such effects occurring with visual sensations. But with tactile sensations there are phenomena that are certainly suggestive of it. Think of how, even when the stimulus remains constant, an itch may grow in intensity, or of how a pain may throb; touch your top lip gently with a bristle and feel how the sensation lingers.[125]

Most stimuli in the real world are as a matter of fact relatively brief, not least because our bodies are continually moving on and our sense organs exploring different parts of the environment. The result, presumably, is that what constitutes the conscious present is largely the immediate sensory *afterglow* of stimuli that have just passed by—the dying-away activity in reverberating sensory loops. And it would follow that the temporal depth and subjective richness of this conscious present is bound to be determined by just how long this activity survives.

What, then, if the fidelity of the loops and hence the lifetime of the activity were to be in some way state dependent: affected for example by general changes in arousal or vigilance, or by mind-altering drugs? It would mean that the depth of the conscious present might to some degree be variable—rather as the depth of sound from a piano may be extended or shortened by the undamping or damping effects of the loud and soft pedals.

I referred earlier to the effects of so-called consciousness-expanding drugs, and to Aldous Huxley's description of his experience under the influence of mescaline: "Visual impressions are greatly intensified. . . . Like the flowers, [the books on my study wall] glowed . . . with brighter colours, a profounder significance. . . ." It seems quite possible that what he is describing is a state of mind where sensory activity continues to reverberate beyond the normal limits and the conscious present lasts quite unusually long. (Perhaps, indeed, this is a state of mind quite "normal" to painters such as Turner.)

By contrast, people sometimes experience states of depression where there is a loss of visual intensity and colors appear flat and washed out: as if, in this case, the life of the sensory activity has been curtailed and the conscious present shrunk.

The most dramatic example of what happens when the reverberating activity is damped right down may be the state of sleep. As a person "drops off" to sleep the conscious present effectively shrinks to nothing and subjective time becomes no more than the shallow stream of physical time.

These suggestions can be pictured, as in the following diagram.

Mescaline:

```
      0      1      2      3      4      5
     1 2 3  2 3 4  3 4 5  4 5 6  5 6 7  6 7 8
     ....4.... ....5.... ....6.... ....7.... ....8.... ....9....
```

Normal:

```
     2 3    3 4    4 5    5 6    6 7    7 8
     ....4.... ....5.... ....6.... ....7.... ....8.... ....9....
```

Depression:

```
      3      4      5      6      7      8
     ....4.... ....5.... ....6.... ....7.... ....8.... ....9....
```

Sleep:

```
     ....4.... ....5.... ....6.... ....7.... ....8.... ....9....
```

$$- - - \rightarrow$$

Subjective Time

....IV.......V.......VI.......VII.......VIII.......IX....

$$- - - \rightarrow$$

Physical Time

· 25 ·

HURRAH!

In the last few pages the terms "conscious" and "consciousness" have entered the discussion again, for the first time since I began this evolutionary history of "what it is to have sensations."

My contention is that consciousness did in fact emerge in evolution as and when these recurrent feedback loops came into being. That is to say, it emerged as and when cerebral sentiments became part of a process that looks forward to its own existence and creates its own extended present outside of physical time.

For human beings (and for other organisms who have reached this same evolutionary level), to "feel a sensation" is to be the author, audience, and enjoyer of the reverberant activity, rolled into one.

Who *says* that consciousness emerged like that? Since I have just said it, evidently I do. But why should anyone accept this say-so? I think they should accept it because, if they accepted the program for solving the mind–body problem that I set out earlier, they will recognize that all the ingredients for explaining consciousness are now in place.

Let me review this program and what has been achieved.

The starting point was the fundamental distinction between sensation and perception. I argued, throughout the first part of the book, that animals have evolved two quite separate ways of representing what happens at the body surface—sensations being affect-laden representations of "what is happening to me," and perceptions being affect-neutral representations of "what is happening out there." This distinction was and remains crucial to everything that has ensued. For, only by insisting on it, have I been able to make my own case: namely, that consciousness, defined as what is *felt* and *present* to the mind, is actually quite limited in scope. Rather than embracing the whole range of higher mental functions (perceptions, images, thoughts, beliefs, and so on), consciousness is uniquely the "having of sensations." And all other mental activities (whether they occur in human beings, nonhuman animals, or even in machines) are outside of consciousness, unfelt and nonpresent to the mind, except and unless they are accompanied by what I called "reminders" of sensation. In short: "I feel, therefore I am" (and, as Milan Kundera has put it, " 'I think, therefore I am' is the statement of an intellectual who underrates toothaches"[126]).

With the problem delimited in this way, the real work of the book could begin—which was to analyze just "what it is to have sensations." In Chapter 17 I examined the salient features of sensations. They include that: "sensations *characteristically* (i) belong to the subject, (ii) are tied to a particular site in his body, (iii) are modality-specific, (iv) are present tense, and moreover (v) are self-characterizing in all these respects." The task, I claimed, was to "explain how these features of sensations could emerge as corollaries of a plausible mechanism in the human brain."

The argument that came next was partly logical, partly biological. I reasoned from first principles that these special features of sensations are and can only be the features of processes that have much in common with bodily activities. It follows that the activity of sensing, which I called "sentition," must have evolved from and still be today an activity that reaches out to do something at the very place where the sensation is felt. In fact every distinguishable sensation in human beings must correspond to a physically different form of bodily activity (either at the real body surface or at a surrogate location on an inner model)—and what it is for someone to feel a particular sensation is just for him to issue whatever "instructions" are required to bring about the appropriate activity.

With this as a basis, I looked back at the evolutionary pedigree of sensation. I showed how present-day sensory activities could have developed step by step from primitive beginnings: starting with a local "wriggle of acceptance or rejection" in response to stimulation at the body surface, later a sensory response mediated by nerves traveling from the body surface to the brain and back again, later still a progressive short-circuiting of this loop by the targeting of the response not on the body surface as such but on the incoming sensory nerve, and eventually the emergence in higher animals of sensory-reverberating feedback loops within the brain.

Thus I have come to a specific hypothesis about the brain mechanism underlying the having of sensations (specific, that is, as to its general logical requirements, not as to its precise physiological basis). This mechanism is physiologically plausible, insofar as it involves nothing more neurophysiologically outlandish than those simple feedback loops. It is clinically plausible, insofar as it is consistent with the evidence for the effects or noneffects of damage to the sensory pathways (phantom limbs, loss of sensation after damage to the sensory cortex, etc.); and, as suggested at the end of the last chapter, it provides too a plausible explanation for changes in the depth of consciousness. Most encouraging of all, it is evolutionarily plausible.

Moreover this mechanism has—or has had at various stages in its history—just about all the phenomenological features that are required. The property of sensations being exclusively one's own follows from sensations being among the activities that "I," my executive self, bring into being. The property of implicating the here and now of an event follows from these sensory activities reaching out to create a physical disturbance at the space-time location that is indicated. The property of having a modality-specific quality follows from the activities associated with different areas of the body surface each having their own "adverbial style." The property of existing for the duration of the subjective present follows from the sensory activities surviving for a non-negligible lifetime even after stimulation ceases. And lastly the property of being self-characterizing follows from the activities looping back to become self-referent instructions for themselves.

Hurrah! Yet, are "all" the ingredients for explaining consciousness now in place? Or is it only "almost all"? Perhaps the claim should be limited to "almost all" until one outstanding issue has been settled.

· 26 ·

HURRAH! —
FOR THE OLD WAYS

As *the review* in the previous chapter showed, I can fairly claim that all the ingredients for explaining consciousness have been in place *at some point* in the course of the discussion—which is to say at some point in the course of evolution. What remains to be shown is that, finally, they are all in place *at one and the same time.*

This problem is not serious overall. In building up the total picture I have, it is true, introduced the various properties of sensations piecemeal—arguing for some of them in relation to one evolutionary stage and for others in relation to a later, revised stage. Nonetheless I have been able to argue that *most* of the already existing features *would* have carried through.

There is no difficulty, for example, in seeing how the essential "belongingness" and "indicativeness" of sensations would have been preserved—since it is clear how, in going from bodily sentiments to cerebral sentiments, the activity in the cerebral loop would still have retained its original indexical properties. However, there *may* be difficulties in seeing how this can work for the equally essential "qualitative character" of sensations—since it is by no means so clear how the activity in the cerebral loop would still have retained its original modality-specific properties.

In telling the story about modal quality earlier in the book, I argued that, as the primitive amoeboid wriggles of acceptance or rejection evolved to become centrally generated sentiments occurring at specialized sensory receptor areas, these sentiments—and the outgoing signals that gave rise to them—would have become distinguished by their "adverbial style." In particular I suggested that the *modality* of the sentiments would have been determined by the *structure* of the epithelium to which they were directed; and the *submodal quality* determined by the nature of the *affective function* they performed there. So that, in the case for example of sensing a sweet smell at the nose, the olfactory quality would have resulted from the fact that the sentiments involved the nasal mucosa, and the sweet quality from the fact that they involved a particular type of positive affect.

The trouble comes in seeing how this story could continue to apply once the sensory responses ceased to reach out to the real body surface and became instead targeted on the surrogate body at the sensory cortex. For it has to be asked why any of the original structural or functional considerations that determined the adverbial style of bodily sentiments would still be at all relevant to cerebral ones.

Presumably the form of sentiments occurring at the cortex can no longer be relevantly determined by the structure of the target, since the different regions of sensory cortex have no structural resemblance to the sensory epithelia from which their input comes, and are in fact all basically alike. There is no reason why, for example, a sentiment occurring at the visual cortex should still be constrained to have the visual style of a sentiment occurring at the retina, or a sentiment occurring at the olfactory cortex to have the olfactory style of a sentiment occurring at the nasal mucosa. Furthermore, since these cerebral sentiments have long ceased to have anything directly to do with bringing about a change in the stimulus environment, there is no reason why the form of sentiments should any longer be relevantly determined by any affective function either.

Indeed it might well be suggested that once cerebral sentiments ceased trading with *bodily* reality, the whole notion of "adverbial style" has become totally redundant: something we would not even have considered if we did not know the history. And in that case, we may be in danger of ending up (like everyone else who has tried it) with a theory of sensations that has ceased to be a theory of sensory quality. To avoid which, I shall have to introduce one final chapter to the story.

"The whole notion of [cerebral sentiments having an] 'adverbial style' has become totally redundant . . . *if* we did not know the

history." But the thing is that we do know the history; or, more to
the point, the thing is that cerebral sentiments do have a history. We
should be able to have recourse therefore to our old friend, evolutionary conservatism.

I am going to digress a bit (and the justification will become
apparent shortly).

In *The Evolution of Designs*[127] the architect Philip Steadman draws
attention to the conservative tendencies shown by human craftsmen
who persist in incorporating elements of past designs into their
contemporary work, long after the original purpose of these elements has been superseded or even totally forgotten. He cites the
example of how, until quite recently, potters in Cyprus "would still
add two blobs of clay to a newly finished jug, without being able to
offer any explanation save that this was the traditional form of
decoration." The explanation, it turns out, is provided "by a comparison with vases up to 2,500 years older found by archaeologists
in the same area. These take the form of finely modelled female
figures. The two protrusions are the lady's breasts."

Design features that were once of practical importance but have
later become mainly if not wholly decorative—and no longer subject to selection on utilitarian grounds—are given the name "skeuomorphs" (from the Greek, "utensil" "form"). Examples are widely
found in clothing (e.g. the buttons on the cuffs of men's coats), in
engineering (e.g. the running boards on early motor cars), and on a
grander scale, in architecture. In classical Greek temples (and their
descendants right up to the present day) many of the decorative
features of the stone buildings hark back to the structural features
of the wooden buildings that preceded them: the dog-tooth Doric
frieze, for example, comes originally from the pattern made by the
exposed ends of timber roof–supporting beams, and the earliest
stone temples even had stone reproductions of the wooden pins.

Craftsmen tend to *copy* preexisting models. And the reasons for
copying are several. Partly it is that copying is easy: the selection or
planning that went into the development of the earlier version is
now inherent in the structure, and the copy can be made without
having to work through this again. Partly it is that copying is safe:
the earlier version did the job required of it, and the copy can be
trusted to do the job at least as well. And partly it is that copying
creates objects that are in tune with what people expect: the earlier
version has set the standard for what the design "ought to" look

like, and the copy ends up looking comfortably familiar. This latter factor is likely to have been especially powerful when, as must have often happened, the old and new versions have coexisted in the same environment and there has been a need to avoid a clash of styles (a stone temple, say, being built next door to a wooden one).

Now, what applies to cultural evolution applies also to biological evolution. In the generation of biological offspring, copying an established pattern is again easy: it requires no redesign work (and basically it can all be left to the existing genes). It is again safe: it provides an assurance that the biological fitness of the offspring will be at least as good as their progenitors. And it is again in tune with preexisting canons: it reduces the risk that one part of the organism will get modernized in a way that clashes with parts that have not changed.

We should expect therefore that living organisms, even as they have evolved new ways of doing things, will have stuck to some of the irrelevant patterns of the past. In other words we should expect to find—and do in fact find— biological skeuomorphs, biological "utensil forms," persisting either as decoration or sometimes just as useless baggage.

The journey of the turtles across the South Atlantic provides one such example. In human beings there are anatomical examples in the vermiform appendix, the wisdom teeth, and the fused vertebrae that form the remnant of the tail; and physiological examples in such oddities as the tendency of our hair to stand on end when we are scared, our liking for the smell of musk, our need to sleep eight hours of the night, and the lunar reproductive cycles of women.

Would it not make sense, then, to argue that the persisting qualitative character of cerebral sentiments—"sentiments that have ceased trading with bodily reality"—is a skeuomorphic feature too?

Consider the following analogy from cultural evolution. There are today a variety of handwritten alphabets in current use: Roman, Greek, Hebrew, Chinese, and so on. Let's suppose (for the sake of the analogy, even if it is not literally true) that the generic style of each alphabet was determined in the past by the physical medium in which the script was written: Roman script was chiseled into stone, Greek scratched with a stylus on wax tablets, He-brew written with a quill pen on papyrus, and Chinese painted with a brush on paper. Furthermore, let's suppose (even if again it is not literally true) that in the past the shape of each individual letter was partly determined by the mouth movements that made the corresponding sound: in Roman script, for example, the letters b and p had their curved section pointing forward because they corresponded to sounds involving a plosive movement of the lips (by contrast, say, with g and d).

Today of course we no longer employ the same writing media, and we no longer mouth the letters as we write; in fact today we have in many contexts given up hand-writing the letters altogether, and resorted to typewriter or printer. Yet we have remained loyal to both features of the ancestral alphabets (even on the computer screen) because to invent a new style of writing would have been difficult, risky, and discordant—and hence any change would simply have been opposed by cultural inertia.

The parallel with sentiments is I hope obvious. Sentiments have continued to keep both the structural and functional components of their adverbial style for these same three reasons operating in biol-ogy. Thus sentiments at the visual cortex, for example, still retain their visual style (as if they were still employing the retinal medium), and furthermore sentiments in response to red light still retain their red style (as if they were still producing a defensive reaction to the stimulus), because any change would have been opposed by biologi-cal inertia.

Two questions must arise, if this is right. First, are sentiments no longer subject to *any* sort of selection on utilitarian grounds: so that their style has become purely "decorative"? Second, has the style of sentiments, in the absence of selection, really remained *totally* un-changed: with the result that the styles of human sentiments still closely resemble those of our distant relations such as monkeys or even frogs or earthworms?

In relation to the first question, we have to remember the *repre-sentational* role that sentiments have always played. Since the very early days, the organism's response to the stimulus has been pro-viding the organism with a mental representation *of* the stimulus:

that is, a representation at the level of sensation of "what is happening to me." And, as we have seen, advanced animals no less than primitive ones still depend on such sensory representations in a variety of ways—not only for the primary purpose of assessing whether what is happening at the body surface is good or bad, but also for secondary purposes in relation to the validation of perception.

We can be sure therefore that there would have continued to be selection to ensure that the *differences* between sentiments were maintained. For example, if responses to light at the retina are to continue to represent the stimulus as light rather than touch, the visual sentiments will have had to remain clearly distinct from tactile sentiments; equally, if responses to red light are to continue to represent the stimulus as red rather than blue, the red sentiments will have had to remain clearly distinct from blue sentiments.

But, given that this distinction might anyway have been maintained just by tradition, why should such an isolating mechanism have been necessary? The reason is that when traditions are handed on merely by copying, without continuing selection pressure, they are always liable to undergo "genetic drift": in other words, small errors in copying accumulate, until the final version may actually have little left in common with the original.

Steadman cites a remarkable example of such drift, occurring in the history of Romano-British coins. Once there was an original gold coin, showing the head of Philip of Macedon crowned by a wreath. But as local copies were made by (slightly careless) British craftsmen: "The Emperor's face rather quickly disappears in the copying, leaving the wreath only. The wreath then undergoes all sorts of remarkable transmogrifications, becoming coarsened in treatment into patterns of rectangles and ovals, turning thence into ears of wheat or barley; while the Emperor's own ear at the centre changes into symmetrical crescent moons, which in their turn attract matching stars." This case is perhaps extreme. But even the Doric frieze has drifted a fair way from a row of wooden beams, and the bumps on the Cypriot pots no longer look *very much* like breasts.

So, this kind of drift might have—indeed we should assume it probably has—occurred with cerebral sentiments. In their case, however, the drift will at least to some degree have been constrained by the need to maintain differences between one sensory representation and another. Selection will have seen to it that the style of visual sentiments for example has never been allowed to become too much like that of tactile ones, or the style of red sentiments too much like that of blue ones.

The same of course applies to written scripts. Over the centuries there has in fact been considerable drift in the precise way alphabets are written. But scripts too have always had a representational role to play: namely the representation of the different sounds of speech. And there has therefore, within each alphabet, been continuing selection pressure to keep the individual letters looking different— preventing bs, for example, ever drifting in the direction of being too much like ds.

Whether there has also been selection to isolate the different alphabets as such is not so obvious. But to make the analogy with sentiments still stronger, let's imagine the following scenario. Suppose that from the beginning different alphabets, as well as being adapted to different writing materials, were employed exclusively for representing different kinds of subject matter: all texts in Roman script perhaps being about optics, all in Greek about acoustics, all in Hebrew about mechanics, all in Chinese about gastronomy. Then, assuming that people would always have benefited from being able to tell at a glance what the subject matter is, there would indeed have been continuing pressure to maintain the generic differences—preventing any set of letters of the Roman alphabet coming to look too much like letters of the Greek one.

This bears directly on the second question of the extent to which the cerebral sentiments of human beings have continued to resemble the sentiments of our far-distant relatives. If there has in fact been drift in the style of sensory responses, but drift limited by this need to maintain the original generic separation, we should expect there to be some degree of resemblance between the sentiments of related species but by no means complete overlap. Just as my handwriting and Cicero's are still authentically "Roman," the visual sentiments of human beings, monkeys, and frogs presumably all still belong to the authentic "visual" tradition. But, even so, just as gothic Roman script has drifted apart from italic Roman script, the visual sentiments of different species may in fact by now each have their own peculiar substyles.

THIS IS THE WAY A FROG SEES

THIS IS THE WAY A RAT SEES

This is the way a monkey sees

𝕿𝖍𝖎𝖘 𝖎𝖘 𝖙𝖍𝖊 𝖜𝖆𝖞 𝖆 𝖍𝖚𝖒𝖆𝖓 𝖘𝖊𝖊𝖘

F i g u r e   1 2

It follows that if a human being were somehow to find himself issuing instructions for monkey visual sentiments instead of human ones, and hence—since this is what it's all about—if he were to *experience* what a monkey does when for example he sees red, then the human being would probably recognize what is happening to him as being a "visual" sensation and even "red": but it might well be a red sensation unlike anything that he has ever sensed before.

But it is not only cross-species comparisons that are of interest. For who knows whether all members of the human species have sentiments that are identical in form? Just as there are small differences between the handwriting of individuals who have learned in the same classroom, so there may quite possibly be small differences between the sentiments of human beings who belong to the same epoch, race, and culture (a possibility that might, one day, open up a whole new field of sensory "graphology"!).

· 27 ·

THE MIND
MADE FLESH

The first task of a theory of consciousness has to be to satisfy the basic scientific and logical criteria that we have been discussing. The theory must describe a physical process in the brain whose properties, at the appropriate level of description, correspond to the properties of felt sensations. With the added ideas of the last chapter, I believe that the ingredients for doing this are, at last, now all in place.

Yet, this is not all that a theory of consciousness must do. For there is no denying that, if the theory is to win the argument in public, it must also satisfy certain other rhetorical or dialectical criteria. In particular it must be responsive to a whole range of supplementary questions that have, over centuries of speculation, moved to the center of both secular and lay discussions of how consciousness fits into the world.

These include the perennial questions about what we can and cannot know about other minds and other brains . . . questions of whether dogs or computers or armchairs are conscious, and how their respective experiences would compare with our own . . . questions such as "What is it like to be a bat?"

They may or may not be good questions; we shall see. But, good or bad, the theory cannot afford to stand aloof from them. At the very least it had better be able to "talk back" in a satisfying way on those issues about which people—rightly or wrongly—feel themselves entitled to be satisfied. Moreover it had better talk persuasively, for they are issues about which most people already have strong—perhaps even unshakable—opinions, albeit without the blessing of any theory whatsoever.

This is not to say that these questions can or should be decided by democratic vote (still less by Dr. Johnson's "I refute it thus"). But it is to say that there is no point in embarking on a losing battle with universally held prejudices. When the question is, for example, "Is a dog conscious?" then we may as well recognize that the only answer that will be publicly sustainable is "Yes," while when the same question is asked about an armchair the only answer will be "No." In short the theory must be talkative and it must talk sense.

What we must do now is to try out some of these questions on my theory—and in the course of it prove, I hope, not only that the theory is exceptionally talkative, but that it does talk an exceptional amount of common sense.

How Far Does the Scope of Consciousness Extend in Nature?

I take it that every reader of this book accepts the premise with which we started back in Chapter 3: that consciousness has temporal and spatial limits in the universe—that there was a time in history when consciousness existed nowhere, and that even today it does not exist everywhere. (The alternative idea, that consciousness has always been inherent in every particle of matter, sometimes called "panpsychism," is one of those superficially attractive ideas that crumble to nothing as soon as they are asked to do any sort of explanatory work.)

It is one thing, however, to accept that there must indeed be limits, and another to make principled suggestions about what those limits are: to suggest why, when, and where consciousness first appeared, and how far and in what contexts the infection has spread. In this respect however the present theory is peculiarly well placed, since it has been systematically developed as a theory of how consciousness emerged in evolution from nonconscious beginnings.

First of all we can conclude that consciousness is strictly tied to *bodies.* To be conscious is essentially to have sensations of "what is happening to me": in other words, of what is happening at the boundary between me and not-me. Without a body there would of course be no such boundary and hence nothing for the subject to be conscious of. This means for example we can entirely rule out the possibility of consciousness in incorporeal entities such as (the admittedly unlikely case of) fields of force, numbers, sound waves, rainbows, universities, pop songs, telephone networks, or immaterial souls or ghosts. We can rule out material entities that, even though they happen to be bounded, have no intrinsic boundaries, such as clouds of interstellar dust, mud flats or snowstorms, and also collective entities made up of separate bounded individuals such as pairs of twins, swarms of bees, or the human species as a whole. For what it is worth we can equally rule out the Universe in its totality, or God in His—since neither has a boundary at which anything can happen (what, in His infinitude, could God feel happening *to* Him?).

Second, we can conclude that consciousness is tied to *self-interested* bodies. Sensations are sensory activities that (in their origins at least) have to do with what is "good or bad." Without self-interest there can be no such evaluation of anything as good or bad and hence no possibility of a response to stimulation having this affective dimension to it. This means we can further rule out the possibility of consciousness in all those corporeal entities that do have boundaries and may indeed respond to what happens at these boundaries, but basically do not care about what happens to them. We can rule out icebergs or rubber balls or pocket watches or the moon. In fact, in the natural non-man-made world, we can rule out all but *living* entities, since none others have an intrinsic interest in their own survival and in none others does the stimulation matter to them.

Third, we can conclude that consciousness is tied to a very special group of living entities, namely those animals that have evolved beyond the stage of a simple sensory response to the critical point where the response has become part of a reactivating loop with a significant *lifetime.* Sensations are intentional activities that persist for an extended moment in subjective time. Without the activity *existing* in this way, the conscious present could only be stillborn and hence the organism could no more be consciously aware of what is happening to it or how it is responding than we ourselves are when we are asleep. This means we can rule out the possibility of consciousness in all those organisms that are still at the stage where the sensory response remains a bodily activity occurring at the body surface rather than at a surrogate location in the brain—and hence in which the loop is too long and too noisy to sustain reverberant activity. We can rule out amoebae, worms, fleas . . .

Earlier in the book I hedged my bets on this. When discussing the case of earthworms responding to light, in Chapter 5, I wrote: "But it is at least arguable that [the way the worm is reacting to the stimulus] should be counted as visual sensation . . . provided we put aside any worries we may have about whether worms are conscious." But now, having reached the point where worrying about whether worms are conscious is exactly where we are at, we can recognize that even if it makes sense to talk of the worm disliking what is happening, it probably does not make sense to talk of it as *feeling* the sensation in the conscious present. In fact it probably does not make sense to talk of any animal that lacks a sensory projection area in the brain as doing this: for what is required is a short high-fidelity loop of the kind that probably occurs only in the cerebral cortex of animals such as ourselves.

Too little is known at present about the anatomy of other species' nervous systems (or indeed of our own) to decide for sure which other animals do have brains that in this respect resemble ours. There is no reason to believe that human beings alone have reached the requisite stage of brain development. But if we are being cautious we should probably think of it as being limited to higher vertebrates such as mammals and birds, although not necessarily all of these.

One thing of which we can be sure is that wherever and whenever in the animal kingdom consciousness has in fact emerged, it will not have been a gradual process. Liberal philosophers, opposed to the idea of any great discontinuities in nature, have sometimes suggested that consciousness has arisen slowly by degrees, with some animals being a "little bit conscious," others more so. But this, according to the theory, is something we can definitely rule out. For consciousness will not have arisen unless and until the activity in the feedback loop took off as reverberant activity: and feedback loops typically have all-or-nothing properties—either they support reverberant activity with a significant lifetime or else the activity dies out at once. Hence, we may guess that, as the sensory loops grew shorter in the course of evolution and their fidelity increased, there must have been a threshold where consciousness quite suddenly emerged—just as there is a threshold that we ourselves cross in going from sleep to wakefulness.

"B.c."—"before consciousness"—sensory responses had no *temporal existence*. But then, as another book says, at some crucial point in history "The Word was made flesh": and it stands to reason that there must have been a comparable Christmas in the evolution of sentition.

What Does This Say About Extraterrestrial Life, or About Artificial Life on Earth—About the Possibility of Man-Made Machines Being Conscious?

Nothing yet said in relation to the theory, nor anything I want to say, would limit consciousness to life on earth. If living organisms have in fact evolved on one of the other half billion or so planets in our galaxy that have a suitable environment for supporting carbon-based organic chemistry, then there is every likelihood that on some of these planets there are in fact creatures that are now conscious for the same historical reasons that we are.

Nor has anything yet said limited consciousness to life based on carbon atoms rather than on silicon atoms or whatever else. According to the theory it is—in the language of computer programmers—the properties of the software rather than the hardware that are crucial: that is, the logical properties of the reverberating circuits, rather than the fact that they are made from nerve cells or that the nerve cells have a particular chemical structure. A silicon-based living organism, for example, might quite well have evolved to have a brain containing circuits with exactly the same logical properties as those we know. And, according to the theory, it too would then be capable of feeling sensations and living in the conscious present.

Hence, if living organisms have in fact evolved on one of the myriad additional planets that might be able to support another kind of organic chemistry, then there is every likelihood that there are conscious creatures on these planets too.

But if conscious creatures made of nonstandard biological materials could live on a far-distant planet, then perhaps they could live on earth as well. And if they have not actually evolved on earth, maybe they could in principle be *manufactured* on earth by human beings. Of course no human engineer would want to be (or be capable of) working in the way that nature does with living tissues such as flesh, bone, nerve cells, skin. But, given that it is the software rather than the hardware that is important, maybe a perfectly good robot version could be assembled from more manageable components such as copper wire, rectifiers, semiconductors, photodiodes, plastic membranes, and so on. In other words, human engineers might be able to manufacture robots, that with their artificial brains, artificial sentiments, and artificial reverberating sensory activity would be artificially conscious—which is to say actually plain conscious.

Maybe it could be so in principle; but there are reasons for thinking it could absolutely *not* be so in practice. I am not talking about the trivial case of an engineer blindly reproducing every jot, tittle, and synapse of a conscious animal's brain, and so ending up with a carbon copy (!) that would by definition have all the same functional properties as the animal. I am talking about the case of building a conscious robot from scratch on the basis of theoretical design principles, knowing what kind of biological and logical requirements must be met. And the reason why this could almost certainly not be done in practice is that there would be no way of re-creating the natural *historical traditions* that have given the activity occurring in natural brains the peculiar *modal quality* of consciousness.

It is true that a robot could be engineered to have something equivalent to a biological "body" with something equivalent to biological "interests," so that it would have at least the possibility of representing and even caring about "what is happening to me." The robot could also, presumably, be engineered to have sensory responses, and these responses could be made to end at a sensory projection area in the brain and become part of a closed loop, so that it would have the possibility of being the author, audience, and, yes, even the enjoyer of the resultant reverberating activity. But all this would not end up imbuing the robot with human-like sensory consciousness, unless the activity in the loops were also to have the right adverbial character. And what would make it extremely difficult to design-in this crucial adverbial character is that the form of naturally occurring sentiments in human beings is, as we have seen, largely a historical accident—a skeuomorphic feature—and not designed-in at all.

The whole point about skeuomorphic features is that they no longer make any "design sense." The engineer, setting out to make a conscious robot, might of course get it right by mere luck: but this would be about as likely as if, setting out to make a well-designed clay pot, he ended up making a pot with knobs on it, or setting out to make a writing machine he ended up making a machine that wrote with a Roman hand. Indeed the only way—other than carbon-copying—of rediscovering the crucial adverbial characteristics of sentiments might be to simulate the whole process of natural evolution that put them there in animals like us to start with. But we have known all along that consciousness can be created by natural evolution. It is the possibility of doing it from a drawing board that is being called in question.

This is not just an *ad hoc*, superficial objection to the idea of artificial consciousness. It is a reason for saying that no rational design process, working from first principles, would ever be likely to succeed. What the engineer is up against is rather like a design equivalent of Gödel's theorem in mathematics. Gödel's theorem states that any system of arithmetic is bound to have properties that cannot be deduced from the axioms: there will be true arithmetical statements, so-called Gödel sentences, that cannot be demonstrated to be either true or false. By analogy (not of course a strict analogy) any naturally occurring biological system will have properties that cannot be deduced from considerations of its contemporary functions: there will be true facts about it that cannot be captured by any design-driven attempt to re-create it.

These biological Gödel sentences might often be of no significance. But in the case of consciousness, they are going to be decisive: making the difference between a conscious robot and a robot whose consciousness essentially lacks conscious quality—which is to say plain unconscious.

WITH THOSE OTHER ANIMALS
THAT *ARE* CONSCIOUS, WHAT KIND OF EVIDENCE
CAN WE EXPECT TO GET OF IT?

The one thing that is not being called in question is that, even if consciousness is unlikely to occur in man-made robots, it certainly does occur in all other human beings, and probably in a fair number of nonhuman animals, both on earth and possibly elsewhere.

Among animals on earth it is of course only human beings that are likely to be able to affirm their consciousness in public, since the only obvious means we have of communicating with others about consciousness is language. It is a fact of life that we cannot hold a conversation about conscious feelings—the kind of conversation that I had with Lily—with a chimpanzee, dog, or magpie (and probably not with an extraterrestrial either, unless he spoke a language we could comprehend). But we can, and from time to time do, hold such conversations with a variety of other human beings. Indeed we can, as I did in Chapter 17, go further and lay on the table the results of particular introspective observations about the special features of sensations—their indicativeness, modal quality, existence, and so on—and thereby seek the assent of other human beings: "Yes, I understand what you are talking about, and yes, the same is true for me." Assuming we get this assent, we have as good evidence as we can ask for that other human beings are in fact members of the same conscious club as we ourselves.

The fact that we cannot do this with other species is a pity. But that's life: and life does impose contingent limits on what we can get evidence about—without those limits being necessarily limits on what is actually the case. The fact for example that we cannot see the other side of the moon from where we happen to be situated does not mean it is not there; and likewise the fact that we cannot confirm in conversation that dogs are conscious does not mean that they are not.

To return however to man-made robots. In the case of robots there has been a tradition of philosophical skepticism that starts by putting the problem just the other way around: asking not how could we know if a robot is conscious (if it is), but how could we know that it is not conscious (if it is not). Following on from discussions of the so-called Turing Test, it has been suggested in all seriousness that, if for example an unconscious robot were programmed to answer questions about consciousness just like a human being does, we might actually be duped into thinking that it was in fact conscious.[128] This unconscious robot, when invited to respond to our observations about sensations in Chapter 17, would also say, "Yes, I—the robot—understand what you are talking about, and yes, the same is also true for me." And hence it might be suggested that, to be consistent with what I have just said about testing for consciousness in other living organisms, we would have to swallow our reservations and at least provisionally welcome the robot to the consciousness club as well.

This, however, is asking too much of consistency. It might in fact be perfectly open to us—without being inconsistent, only sensible—to claim that a test appropriate to another living organism could not be expected to yield reliable results if applied to an entity that is *operated by* or *designed by another conscious being.* A ventriloquist's dummy for example could also pass the conversation test. But in this case, rather than concluding that the dummy was conscious, a more sensible conclusion would obviously be that it was the dummy's operator with whom we were in conversation—and hence the operator not the dummy who had given the evidence of consciousness.

The case of the unconscious man-made robot would be a little different since there would not be a conscious being directly operating him. There would however have been someone responsible for his construction and design. And if the robot were dissimulating so effectively, he could only be doing so because this designer himself knew what kind of answers were required—for we can be sure that a designer who was not himself conscious would not have been able to write a sufficiently convincing program. The sensible assumption would therefore again be that it was this designer with whom we were indirectly in conversation—and hence this designer not the robot who was conscious.

Suppose, however, that we were not permitted to be sensible like this, only stupidly consistent. The situation with the conversation test would still not be too bad. For even though we might end up making the mistake of concluding that the robot was conscious, it would be only half a mistake. The test would have correctly diagnosed the hand of consciousness *somewhere:* if not in the robot itself, then at one remove in the designer. We should, I think, be happy to accept this situation. In a world where we cannot have complete knowledge of the tricks that may be played on us, the fact is we must sometimes get taken for a ride: and that's life too (and not a philosophical disaster).

How Does the Quality of the Experience of Other Conscious Animals Compare with Ours?

If and when other animals are conscious, what they are experiencing is the sensory activity in their own cerebral loops. And, according to the theory, the quality of their sensations will be directly related to the adverbial style of the corresponding sentiments. It should therefore be possible, in principle, to state the conditions under which one animal's experience will be similar to or different from another's.

I raised some of the relevant considerations at the end of the last chapter when discussing how the style of sentiments might have "drifted" in the course of evolution. In the light of that discussion, we should expect that within a single species there will be a large degree of overlap from one individual to another, with only minor "graphological" individual variations. Thus, any other human being's sensation of sweetness, for example, is likely to be highly similar to our own. Between closely related species there will still be overlap, although it may be considerably less because of the greater potential for genetic drift. Even so we should expect there to be at least a generic resemblance: a monkey's sensation of redness, a dog's sensation of pain, or a bear's sensation of sweetness are likely to be at least in the same qualitative league as mine or yours.

Hence when the question is, as it is often put: " 'What is it like to be' some other individual in a particular sensory environment?" we need not be shy of offering an answer. The answer is that what it is like to be another *human being* is probably very much like what it is like to be ourselves in the same environment; and what it is like to be another *closely related animal* is probably pretty much like it. (I am assuming the question of "what it is like" is restricted to the basic quality of the sensation, not to any higher-level "thought" about it: bears and humans, even if they both have similar sensations in tasting honey, need not of course think about honey in like terms at all.)

This answer relies however on one obvious condition: namely that we and the other animal have very similar sense organs. If the other individual with whom we are comparing ourselves were to lack our own sensitivity to a particular form of stimulation or if he were to be sensitive to a kind of stimulation that we are not sensitive to, then what it is like to be him in a particular sensory environment might of course be altogether different from what it is like to be ourselves.

How different? And what would *that* be like? There is clearly no great problem with imagining the experience of another animal whose sensitivity to a particular kind of stimulation is lower than our own, being color-blind, say, or deaf. Nor need there be any very great problem with another animal whose sensitivity is greater than our own, being sensitive to ultraviolet light, say, or ultrasound— provided the sensory modality is one we are familiar with. Within a particular modality the available "adverbial space" for sentiments is presumably limited, and given the need to keep sentiments as distinct as possible, it would make sense if animals evolved to use this space to the full. Hence, if for example an animal can hear sounds of either lower or higher pitch than we ourselves can, we could plausibly assume that the lowest sound he hears has the sensory quality of the lowest sound we can hear and the highest sound that of the highest: in other words that the qualitative range of his sensation is similar to what we already know, even if it happens to cover a different range of stimulation.

What might present a major problem, however, would be if another animal were to be sensitive to a kind of stimulation that lies outside of any sensory modality we know—raising the possibility of this animal feeling sensations of a qualitative type that has never been experienced by any human being. The example that has been most widely discussed among philosophers is that of the echo-locatory sense of bats; but other examples might be provided by the electric sense of lampreys or the thermal sense of pit vipers.

What is it like to be a bat? The bat's case, despite all the attention it has been given, may not be particularly interesting in this respect—since it is far from obvious that bats' echo-location does in fact involve an alien sensory modality. Bats, in their ability to echo-locate, certainly have a *perceptual* ability that is unlike anything that we human beings possess: in other words they have an exceptional ability to use information arriving at their ears to represent "what is happening out there." But this is no reason to believe that they have *sensations* that are unlike those we know: in other words that there is anything exceptional about how they represent "what is happening to me." The sense organ involved in echo-location is, after all, not a generically new sense organ: it is the typical mammalian ear—an ear very much like our own. And when sound waves reach the bat's ear and excite the basilar membrane, the form of the bat's sensory response—the adverbial form of its sentiments—is presumably as much in the auditory tradition as any other mammal's. Hence what it is like for a bat to receive sound at its ears is probably not all that dissimilar to what it is like for us: even when it is echo-locating it is experiencing the returning whistles as nothing more or less exotic than high-pitched auditory sensations.

The case of skin-vision provides a helpful analogy. A human being wearing the skin-vision apparatus described in Chapter 10 also (after a little training) possesses a perceptual ability that most of us do not. Yet he does not acquire any new sensory capacity: when the vibrators tickle the skin of his back, he still represents "what is happening to me" as being tactile in quality. It is true, as we noted, that he may in fact give all his attention to the perceptual channel, and so mask the tactile sensations altogether; and the same could be the case with bats. In the excitement of the moment, when for example chasing prey, bats may not be consciously aware of anything happening at their ears at all. Nevertheless, if they are aware of anything "happening to me," it will be of having an *auditory* experience.

But if bats do not provide an interesting test case of an exotic sensory modality, is there any animal that does? What *would* result in an animal having a sense organ that gives rise to sensations in a modality that we human beings know nothing of? According to the theory, cerebral sentiments have followed in the tradition of the bodily sentiments whose modal style was originally determined by the nature of the sensory epithelium at which they occurred. Hence it is only if an animal possesses a sense organ that has originated quite differently from any human sense organ, from a structurally different kind of sensory epithelium, that it will now have cerebral sentiments whose modal style is unlike any of our own. That is to say: only if the animal has a sense organ that *does not share a common ancestry* with any of ours. Among higher vertebrates, however, there are no examples of such totally alien sense organs. All human sense organs and all those of other vertebrates have evolved from the same set that were already present in the ancestral fishes from which we have all come. This is true even of highly modified organs such as the heat-sensitive pit organ in the forehead of the viper, or the electric organ in the body of the lamprey.

We can therefore conclude that there are probably *no* totally-unknown-to-us sensory modalities, at least in vertebrates. In invertebrates, admittedly, there might be. But we have already concluded that invertebrates, without a sensory cortex in the brain, are unlikely to be conscious anyway.

Supposing That We Ourselves Had Never Experienced a Particular Modality of Sensation, Where Would That Leave Us?

When I wrote that there are probably no "totally-unknown-to-us" sensory modalities, I meant of course "us" as normal human beings, in possession of the normal range of human sense organs and with appropriate experience of their use. If a human being were to be missing one or more of these organs—if he were to have been born blind or deaf, for instance—his position would obviously be rather different.

Would there be no way he could find out—perhaps at second hand—what it is like to experience the missing sensory modality? Common sense suggests not, and so does the theory I have been putting forward.

Since sensations always relate to what is happening to "me," then to know what it is like to feel a particular sensation has to be to know what it is like for "myself." And since for myself to feel a sensation in a particular modality is to be the *author* of sentiments with the corresponding modal quality, only someone who is in a position to be such an *author* can know what it is like for himself. But someone who, for example, has no eyes and no visual cortex, cannot possibly be in a position to be the author of visual sentiments. *Ergo,* he cannot know what it is like to have visual sensations.

It is the intentionality of sensations, the subject's essential part in *issuing instructions* for the sentiments, that makes it impossible for anyone to enter in at second hand unless he has the relevant equipment to create the corresponding sentiments himself. Oscar Wilde, on hearing of a witty remark made by someone else, commented to a companion: "I wish I had said that." His companion replied: "Don't worry: you will, Oscar, you will." A fair prediction, since Wilde did (notoriously) have the right equipment for making or repeating the full spectrum of witticisms. But suppose that Wilde had had a brain lesion that rendered him partially aphasic, so that he selectively lacked the ability to utter this particular genre of remark. Then the only reply his companion could honestly have given would have been: "You won't, Oscar, you won't."

Consider, as a thought experiment, the hypothetical case of a brain scientist called Marian (a related case, though not quite this one, has been discussed by Frank Jackson[129]). Marian is a physiologist who studies the visual system of other human beings but is herself totally blind because she has no visual pathways in her brain. Through her research, employing her other sense organs, Marian gets to know everything it is possible to know *from the outside* about what goes on in another person's brain when this person for example has a red sensation. That is to say (since we may assume she has confirmed the fact of sentiments) she knows everything there is to know from the outside about visual sentiments, including the exact adverbial style of the sentiment associated with seeing red. The question then arises: does this mean that Marian knows for herself what it is like to have a red visual sensation? On the basis of my theory we can surely answer: No. For, even if Marian knows everything there is to know about sentiments from the outside, she still does not know what it is like to be the author of them. And since she lacks the cerebral equipment to be the author, this is something she could never ever know.

Certain philosophers have gotten very bothered about cases such as Marian's. Some have seen a deep mystery in her inability to enter into the sensations of the subjects she studies so exhaustively; others have claimed that if she cannot know what it is like for them to have sensations, this can only mean there is nothing special to be known—indeed that the whole notion of sensations is a miasma. Yet there is, I suggest, no more need to be bothered about blind Marian's disability than there would be about aphasic Oscar Wilde's. Wilde (let's suppose) is unable to utter a certain kind of joke. That is his tragedy. Marian is unable to utter a certain modality of sentiment. That's hers.

The difference between my theory and any that has preceded it is that it makes the feeling of sensations equivalent to an *action* by the subject. "Feeling," according to the theory, is a kind of "doing." Even if it were true that a person could in principle learn everything there is to learn about the outside world, and so acquire total knowledge of what is objectively knowable, it would hardly be surprising if there are limits to what an individual person can *do*, and so limits to what he or she can subjectively *feel*.

· 28 ·

WATER AND WINE

I *warned in the* Preface that the solution to the problem of consciousness might turn out to be boringly straightforward. Now it comes to it, I think the warning was unnecessary. Conscious feeling, it has emerged, is a remarkable kind of intentional doing. Feelings enter consciousness, not as events that happen *to us* but as *activities* that we ourselves engender and participate in—activities that loop back on themselves to create the thick moment of the subjective present.

The proffered solution is not boring and it is certainly not straightforward. Even so, there are bound to be critics (Colin McGinn would surely be among them) who are going to find it disappointingly mechanistic and unmysterious—lacking a certain "ils ne savent quoi." "Is that all?" they might object. "All we seem to have ended up with is a string of nerve impulses, or information, flowing around a physical circuit in the brain: and—whatever its pedigree, however good its logical and psychological credentials—this hardly seems good enough to underlie consciousness in all its glory. Call it a special sort of 'doing' if you like, call it being the 'author' of recirculating sensory activity. Still, is that *all?* Is consciousness *just* that?"

"The difficulty here is one of principle," Colin McGinn wrote. "We have no understanding of how consciousness could emerge from an aggregation of non-conscious elements such as computational devices; so the properties of these devices cannot *explain* how consciousness comes about or what it is."[130] But it is not only McGinn. I quoted Ray Jackendoff at the beginning of the book: "I find it every bit as incoherent to speak of conscious experience as a flow of information as to speak of it as a collection of neural firings." And the same worries are widely in evidence elsewhere. Thomas Nagel, for example: "We have at present no conception of how a single event or thing could have both physiological and phenomenological properties, or how if it did they might be related."[131] Or Robert van Gulick: "We simply have at present no theories, functionalist or otherwise, that explain how a physical system can have a phenomenal life."[132] Or T. H. Huxley: "How it is that anything so remarkable as a state of consciousness comes about as a result of irritating nervous tissue, is just as unaccountable as the appearance of the Djin, when Aladdin rubbed his lamp."[133]

There could, I admit, still be grounds for anxiety on this score. And yet I do not think they are any longer so serious as these people seem to be suggesting. Indeed I suspect their continuing despondency is, partly at least, a hangover from earlier days when the theories of consciousness on offer had not come anywhere near—and certainly not so near as we are now—to doing the job required of them.

"Is that all?" Is a human skull just a lump of calcium phosphate, is a flour mill just shafts and cogs and wheels, is Hamlet's body just a quintessence of dust? Is water just hydrogen and oxygen, is hydrogen just a proton with a single circulating electron, is the electron just a wave function, a mathematical abstraction? Is the answer to the riddle of life, the universe and everything just 42?

In every case the answer that would be expected to a question posed like this would almost certainly be "No": maybe the thing in question is in fact whatever has been stipulated, but it is not just that—that is not all it is, it is not nothing but that.

There is of course nothing in the world that is finally and absolutely "just" what we may have chosen to describe it as—for the simple reason that there is nothing in the world that could not, if we were to choose otherwise, be redescribed from a different point of view. Even the number 42 could if we chose be redescribed: for it happens to be, among many other things, 7 times 6, the age of one of my sisters, the distance in miles from London to Cambridge, and the magic constant of the smallest magic cube (not to mention its continual reappearance in the works of Lewis Carroll—as, for example, in Rule Forty-Two of the Wonderland Legal Code: "All persons more than a mile high to leave the court").

What matters in the end is that questioner and answerer should have the same point of view, the same agenda, and be interested in the same things. When the question is "What is a skull?" an anthropologist will not be satisfied by the answer that satisfies a chemist. When the question is "What is the purpose of existence?" a mystic will want a different answer from a bus driver. A cosmologist who would have no time for the suggestion that the answer to the riddle of life, the universe, and everything is the distance in miles from London to Cambridge, might well be considerably happier with the suggestion that the answer is the magic constant of a magic cube.

Given the variety of people who have had, have now, and will have in the future their different reasons for asking the question "What is consciousness?" there are no doubt a variety of answers that would be likely to prove more or less convincing or congenial. My answer may indeed be less than the complete answer to someone else's question.

Nonetheless we should not give way too readily to those critics who protest "Is that all?" In developing the theory of consciousness as sensory activity, I have explicitly argued for a particular view of what the question means, and come up with a corresponding view of what the answer is. Since I have been explicit about my point of view, we might expect the critics to be explicit about theirs. If this answer is not good enough for them: what else do they want? And, whatever it is they want, or think that they want, are they sure they have not already got it, without realizing it?

Complaining about the inadequacy of theories of consciousness has, as I said, become so habitual among philosophers of a certain disposition, that there is a real danger that they will carry on saying "Is that all?" even when they no longer have anything substantial to complain of. In Chekhov's play The Three Sisters, the heroines spend the whole play sighing about how wonderful it would be if only they could go to Moscow, when the fact is they have more than enough money in their pockets to take the train anytime they choose.

Let me return again to the statement by McGinn I quoted at the very start. "Somehow, we feel, the water of the physical brain is turned into the wine of consciousness, but we draw a total blank on the nature of this conversion. Neural transmissions just seem like the wrong kind of materials with which to bring consciousness into the world. . . . The mind–body problem is the problem of understanding how the miracle is wrought."

It sounded—McGinn of course meant it to sound—like an impossible task. Yet here we are. Working entirely with the natural syrup (why call it water?) of the physical brain, we have overseen a fermentation process that looks remarkably like vinification. Even if the product lacks the refinement of a *grand cru*, it is a fairly impressive *ordinaire*. The appellation and vintage are certainly respectable (a *vin du terroir*, going back some several hundred million years). The finished product has plenty of body, a nice balance of positive and negative affect, a rich qualitative color, a strong hint of subjectivity, an aftertaste of intentionality, even the suggestion of latent objective phenomenology. Moreover, as an accompaniment to a main course of philosophy, it is unusually accommodating and responsive, complementing a range of both the traditional and *nouvelle* dishes—other-mind pie, bat soup, pickled Turing, fricassee of robot—without being so heady as to lead people to say things they regret.

If McGinn still wants to deny that it is the wine of consciousness, let him taste it and say what is missing.

I confess that I too have sometimes been subject to the "Is that all?" malaise, and would in the past have been all too ready to join McGinn in worrying about what *else* a theory of consciousness ought to do. But, like a sickness that, when once thrown off, seems to have belonged to a different person altogether, the worries no longer seem like *my* problem. Indeed, though there are many details to be worked out, I would now say that neural transmissions seem to me like just the right kind of materials to bring consciousness into the world. And, if I draw a total blank on something, it is not so much on how the conversion takes place, as on whatever made it look like an impossible miracle to start with.

Yet, I tell a lie. For I can guess what the problem may still be. The theory I have been developing, for all its special features, is basically a version of an "identity" theory, and, at that, a "functionalist" identity theory. And it might still be argued that it is no more *metaphysically complete* than any other theory of this type.

Identity theories, to the effect that X *is* Y, maintain that whatever is described by one term of the identity, X, is the very same thing as whatever is described by the other term, Y; not that the two terms themselves are the same description (which of course, except in trivial cases, they are not), but that they designate or pick out the same thing in the world. And functionalist identity theories maintain, furthermore, that one of the terms of the identity can be well described purely as a logical operation, relating causes to effects, or inputs to outputs, without reference to the material structure involved in bringing about the operation.

Thus, when we suggest that consciousness *is* the activity of being the author of reverberating cerebral sentiments, we are suggesting not only that what is designated by the term "consciousness" is the very same thing as what is designated by the term "being the author of reverberating cerebral sentiments," but that the latter term is to be considered as a logical operation that is independent of what neural or other structures are involved.

Now, even though I would claim that this theory of consciousness does not suffer from the apparent defects of previous functionalist theories that have patently made the wrong identifications, it might still be argued that it cannot be the complete explanation. For however well it succeeds in establishing the *terms* of the identity, it does not explain the underlying *reason* for the identity. That is to say, however well it succeeds at a scientific level in answering the question "What formal operation in the brain is identical to consciousness?" it does not deal with the deeper question "Why is this operation identical to consciousness?"

The latter question may sound like a classic example of a silly question. But I would accept that, possibly, it need not be. For, as Saul Kripke in particular has insistently argued,[134] there may be two kinds of identity, one of which is much more open to question than the other.

On the one hand, there are those *necessary* identities, that are in the last analysis tautologically true, and so must hold true in all possible circumstances in all possible worlds. For example: the number 42 is the product of the numbers 7 and 6; alcohol is what you get by oxidizing sugar; monochromatic yellow light is electromagnetic radiation with a wavelength of 580 nanometers; parallel lines are lines that run in the same direction; a dollar is worth 100 cents. In all these cases the two terms, when and if we come to understand them, turn out to be such that it would be a contradiction to deny that they refer to the same thing. That does not mean that everyone need be immediately aware of the identity, or that we do not have to work to prove that it is the case. It is to say, however, that when and if we have proved it, we have explained it, and it would indeed be silly to ask the further question "Why?"

On the other hand there are those *contingent* identities, that only happen to be true because things are arranged that way in the world in which we live, and so need *not* be true in all possible worlds or in all circumstances. For example, 42 is the number of the bus that takes me home (but not if I happened to live in Paris); alcohol is what is produced when grapes are left to rot (but not if conditions are too cold); the color people see when yellow light reaches their eyes is the color they see when a mixture of red light and green light reaches their eyes (but not if they do not have trichromatic color vision); parallel lines are lines that never meet (but not if you are doing your geometry on the surface of a sphere); a dollar is worth eight rubles (but not on the black market). In all these latter cases the two terms happen to pick out the same thing in a particular world; but it would certainly not be a contradiction to deny that they must do so in some other world. Thus even when we have discovered the identity, we may not have fully explained it, and so it would not be silly to ask the further question "Why?"—why, that is, it holds in one world rather than another.

Now, in the case of consciousness, which kind of identity are we dealing with? When we say that to be conscious is to be the author of reverberating cerebral sentiments, is this an identity that holds true everywhere imaginable: so that, for example, anyone in any possible world who was doing what we are doing when we authorize pain sentiments would be consciously feeling the same pain we are? Or is it an identity that holds only in a restricted world or set of worlds: so that a creature on another planet or in another universe could be issuing functionally identical pain sentiments and not be feeling pain at all? And if the identity is contingent and not necessary, then what is so special about the worlds in which it does hold, as against those where it does not? What quirk of God or nature could be making it *so* in one case, *not so* in another?

People in the past have indeed been prepared to accept that consciousness accompanies brain events only in very special circumstances. Descartes, in particular, maintained that the identity holds for the brains of human beings but not for those of any other animals, and believed the reason "why" to be none other than that God arranged it that way. Yet even if few philosophers now go for *this* kind of contingency and most accept that the identity—if it holds at all—holds pretty widely, a good many would still insist that this need not mean it holds universally and that there is probably *some kind* of unknown (or even unknowable?) contingency involved. For they simply cannot find it in themselves to allow that it could be the case that particular conscious sensations are necessarily identical to particular brain states: that, for example, it would be logically impossible for someone to be the author of recirculating pain sentiments without feeling a pain sensation. And their reason is (at least this is Saul Kripke's reason) that they can, or so they say, perfectly well *imagine* a world—it may not be our world, but so what—in which exactly the same functional state could exist in a being who was in fact not conscious of pain. Since there is no denying that an imaginary world is a possible world, this must surely be sufficient to sustain the argument against necessity.

I would have to agree that, if people were right about their ability to imagine a world where the identity we have been discussing does not hold, then it would indeed be both reasonable and important to pursue the question of why it holds in our world. Just as, if someone were right about their ability to imagine a world in which 42 does not equal 7×6, it would be reasonable and important to ask why 42 is equal to 7×6 in this particular world. But the question is: could they be right about their ability to imagine this—in either case?

In the case $42 = 7 \times 6$, there would be strong grounds for saying that they could not be right. There is admittedly nothing to stop people trying to imagine whatever they like. They could perhaps even find it a useful spiritual exercise to try to imagine that 42 does not equal 7×6 ... or that there is life after death, or that they can hear the sound of one hand clapping, or that their heads are made of mustard. But it is one thing to try and another to succeed. And if someone claimed that they actually were imagining that 42 does not equal 7 × 6, we ought not to be too impressed. Perhaps, being charitable, we might suppose that they had made an honest mistake or were under an illusion; or, not being so charitable, that they simply did not know what they were talking about. For 42 = 7 × 6 really is a necessary identity. And while someone might perhaps be able to imagine some superficially similar identity ceasing to hold, they could not be imagining *this* one not holding.

Then, should we be more impressed with someone who claimed that they can imagine a creature being the author of reverberating pain sentiments without feeling pain? I am inclined to say the cases are exactly parallel, and for the same reason. If someone did claim to imagine a world in which this relationship ceases to hold, we ought to conclude either that he is making a mistake or else that he has failed to grasp the theory. And, while someone might be able to imagine some other version of identity theory ceasing to hold, they could not be imagining this theory not holding. For I suspect that this particular identity is in reality a necessary identity.

Kripke, admittedly, reaches just the opposite conclusion. But then the difference between us is that, for Kripke, any argument for an identity theory that purports to show that "these things we think we can imagine are not in fact things we can imagine . . . would have to be a deeper and subtler argument than I can fathom and subtler than has ever appeared in any materialist literature that I have read." Though I hesitate to say it, the difference between us may well be that Kripke has not been with us for the last ten chapters.

The trouble is the waters have been considerably muddied by *bad* theories: theories involving claims about identity that do not hold even in the world we actually live in, let alone in all possible worlds.

I had occasion recently to look up the article on "Conjuring" in the 1929 edition of the *Encyclopaedia Britannica*,[135] and chanced on the following entry under "Consciousness": "One theory holds that each atom of the physical body possesses an inherent attribute of consciousness. . . . A second theory assumes that there exist, in the brain, special nerve cells capable of producing consciousness whenever activated. . . . The psychonic theory [which the author of this entry, W. M. Marston, clearly favored] suggests that consciousness occurs each time any unit of junctional tissue between individual neurones is energized. Units of junctional tissue are called *psychons,* and each *psychonic impulse* is regarded as a single unit of physical consciousness. This theory is now under experimental investigation.''

History does not relate what became of the experimental investigation of this remarkable theory. But if a philosopher were now to take the psychonic theory as his model, and insist that he can perfectly well imagine a world in which psychonic impulses could occur in, for example, the tail of a lobster without consciousness being present, I would be the last to dispute it. Indeed, notwithstanding the outcome of a hundred experimental investigations, I cannot imagine any world in which the theory holds at all.

But this is not the theory I have been proposing. And what I do dispute is that anyone who understands my theory could imagine *this* theory not holding universally.

The trouble with the psychonic theory is that there is absolutely nothing about it that feels right, that strikes any sort of chord. The theory was not (I assume) driven by any considerations of what the experience of consciousness actually amounts to, at the level of phenomenology or language or behavior: and so, when it comes to it, the theory cannot give the experience back. By contrast my theory began with the salient properties of consciousness and systematically fed them into the identity: it therefore can and, when required to, does give them back.

The result is that to imagine a creature anywhere at any time doing what we do when we play host to reverberating pain sentiments—that is, to imagine this creature being the *author* of the sensory activity, and living in the *extended present* of sentition—just is (if we are successful) to imagine this creature being conscious of a pain sensation. The body side of the equation leaves nothing undesignated that is designated by the conscious side, and vice versa.

But is that all? I do not know what else there is to say. "The art of life," Henry Thoreau remarked, "of a poet's life, is, not having anything to say, to say *something.*" But the wiser course, if you are not a poet, is to stop.

· 29 ·

BEING AND
NOTHINGNESS

I *stopped: but* I stopped on too
downbeat a note to end such a
remarkable history.

"A History of the Mind" has been, as I said it would be, only a
partial history of a part of what constitutes the mind. It has been,
nonetheless, the history of how, over the last four billion years, the
minds of animals have totally transformed the status of the universe
in which they live.

Let me end it with the story of a particular event—a patch of
sunlight arriving at the surface of our planet.

Long long ago, before there was any life on earth, rays of light from
the late-afternoon sun fell on the surface of a shallow rock pool at
the sea edge, passed through the water and were absorbed by a
pebble on the pool's bottom. The pebble like everything else in
nature was insentient. And so the sun set on a world devoid of
meaning, where nothing existed *as* anything *for* anyone at all.

Life began to evolve in these pools; and soon enough the seas were
teeming with tiny self-interested organisms. In this same rock pool
there came to live a protozoan that fed on the debris that floated
near the surface. Now, when sunlight fell on the pool, a little of
it—ever so little—was absorbed at this protozoan's boundary. But
the protozoan, unlike the pebble, was sensitive to light. At midday
it was at risk of being damaged by ultraviolet rays and so it wriggled
away; but as the sun went down it could safely float back up to the

surface. The protozoan was representing the sunlight—through its actions—as an event of significance "to me."

Evolution progressed, and a fish came to inhabit the same pool. The fish lived in a bed of weeds and darted out from this shadowy environment to catch its prey. Light mattered to the fish as well: its optimal environment was the zone where the weeds ended and the clear water began. The fish still retained a light-sensitive skin, and by comparing the stimulation at different parts of its body it was able to adjust its position to keep its tail dark, head light. But the fish had also developed an image-forming eye, and it had taken advantage of the image at the retina to develop a new faculty of vision: the image was being interpreted not just as evidence of the direction light was coming from, but as a sign of what was happening "out there." If the fish had looked up to the sky it might even have perceived a shimmering red disc beyond the pool; but the wind was blowing and the ripples washed this far world out of sight.

Close to where this rock pool once existed, there now stands the city of Cambridge. And living there now is me. Looking from my window at this moment I can see the sun setting over the western horizon. In the tradition of my ancestors I am representing the light arriving at my retina both as a circular patch of redness happening to me and as a fiery orb existing in the galaxy out there. But something else has followed in the course of evolution: the seeming miracle of consciousness. I am now living in the present tense of the sensations that "I" bring into being. I am encapsulating my own response to the sun's image as an activity of which "I" am the author. I have, as it were, taken a loop out of the thin rope of physical time, lassoed the sun—and made it, momentarily, mine.

I do not care to estimate what absolute value we should attach to this transformation of the universe, or how far we should prize some aspects of it above others. Thomas Gray, in his "Elegy," spoke where philosophers might be wise to hold their tongues:

> Full many a gem of purest ray serene
> The dark unfathomed caves of ocean bear;
> Full many a flower is born to blush unseen,
> And waste its sweetness on the desert air.[136]

But it is not only sentimentalists like Gray who would consider a world that passes unrepresented by a mind to be a world whose destiny is sadly unfulfilled. If the question is, "Who is to say what 'waste' is?" I think that we all know.

It is true that any sort of minding is an existentially significant event. The amoeba that turns away from the light, the frog that snaps at a fly, the man whose pupils contract while he sleeps, the blindsight patient who reaches out to grasp a ball—are all doing something that gives the world a dash of meaning that otherwise it would not have.

Yet in the end it is *conscious* minding that has added the new dimension of semantic depth. For it is consciousness, with its power to make the vanishing instant of physical time live on as the felt moment of sensation, that makes it LIKE SOMETHING TO BE OUR-SELVES—and so sweetens and enriches the being of the external world FOR US.

A seeming miracle? No, as close to a real miracle as anything that ever happened. The twist may be that it takes only a relatively simple scientific theory to explain it.

NOTES

1 Nicholas Humphrey, *Consciousness Regained* (Oxford: Oxford University Press, 1983).

2 Nicholas Humphrey, *The Inner Eye* (London: Faber and Faber, 1986).

3 William Calvin, *The Cerebral Symphony* (New York: Bantam Books, 1990), p. 3.

4 Roger Penrose, *The Emperor's New Mind* (Oxford: Oxford University Press, 1989), p. 412.

5 Douglas Adams, *The Hitchhiker's Guide to the Galaxy* (London: Pan Books, 1978).

6 Samuel Coleridge (1801), quoted by Richard Holmes, *Coleridge* (London: Hodder and Stoughton, 1989), p. 300.

7 John Bunyan (1678), *The Pilgrim's Progress*, part 2 (London: Collins, 1910).

8 William Drummond of Hawthornden (1623), *The Cypresse Grove*, quoted by John Hadfield, *A Book of Beauty* (London: Edward Hulton, 1952), p. 183.

9 Duncan MacDougall (1907), quoted by James E. Alcock, *Parapsychology: Science or Magic?* (Oxford: Pergamon, 1981), p. 11.

10 René Descartes (1641), *Meditations on First Philosophy*, Second Meditation, 24, trans. John Cottingham (Cambridge: Cambridge University Press, 1986).

11 Samuel Johnson (1759), *The History of Rasselas, Prince of Abyssinia*, ed. J.P. Hardy (Oxford: Oxford University Press, 1988).

12 Colin McGinn, "Can We Solve the Mind-Body Problem?," *Mind* 98 (1989), 349–66.

13 Gottfried Leibniz (1714), *Monadology*, section 17, quoted by C.L. Hardin, *Color for Philosophers* (Indianapolis: Hackett, 1988), p. 134.

14 William Lycan, *Consciousness* (Cambridge, Massachusetts: MIT Press, 1987), p. 37.

15 Colin McGinn, "Could a Machine Be Conscious?," in *Mindwaves*, ed. Colin Blakemore and Susan Greenfield, (Oxford: Blackwell, 1987), p. 287.

16 Ray Jackendoff, *Consciousness and the Computational Mind* (Cambridge, Massachusetts: MIT Press, 1987), p. 18.

17 T.S. Eliot (1917), "The Love Song of J. Alfred Prufrock," *Collected Poems 1909–1962* (London: Faber and Faber, 1974).

18 Plato, *The Republic*, Book 8, 546, trans. H.P.D. Lee (Harmondsworth: Penguin, 1955).

19 Thomas Nagel, "What Is It Like to Be a Bat?," *Philosophical Review* 82 (1974).

20 In *The Mind's I*, French translation, *Vues de l'Esprit*, ed. D. Hofstadter and D.C. Dennett (Paris: InterEditions, 1985).

21 George Eliot, *Journal*, 20th July 1856, in *George Eliot's Life as Related in Her Letters and Journals*, ed. J.W. Croft (Edinburgh, 1885).

22 George Eliot (1871), *The Mill on the Floss* (London: Folio Society, 1986), p. 9.

23 Stephen J. Gould, in conversation with Colin Tudge, BBC Radio 3, *The Listener*, September 20, 1984, p. 19.

24 John Crook, "The Nature of Conscious Awareness," in *Mindwaves*, ed. Blakemore and Greenfield, p. 392.

25 Kathleen V. Wilkes, "—, Yishi, Duh, Um, and Consciousness," in *Consciousness in Contemporary Science*, ed. A.J. Marcel and E. Bisiach (Oxford: Clarendon Press, 1988), p. 38.

26 Anthony J. Marcel, "Phenomenal Experience and Functionalism," in ibid., p. 121.

27 Alan Allport, "What Concept of Consciousness?," in ibid., p. 159.

28 William James, "Does 'Consciousness' Exist?," *Journal of Philosophy, Psychology and Scientific Method* I (1904).

29 Schoolboy (sixth grader) quoted in *The Boston Globe*, January 25, 1988.

30 Maurice Burton, "The Loch Ness Monster: A Reappraisal," New Scientist (1960), 773–75.

31 Peter Scott, cited in "Naming the Loch Ness Monster," *Nature* 258 (1975), 466–68.

32 Pablo Picasso, quoted in *Aesthetics in the Modern World*, ed. Harold Osborne (London: Thames and Hudson, 1968), p. 24.

33 Thomas Reid (1785), *Essays on the Intellectual Powers of Man*, Essay 2, 17, (Cambridge, Massachusetts: MIT Press, 1969).

34 Ernest G. Schachtel, *Metamorphosis* (London: Routledge and Kegan Paul, 1963), p. 83.

35 E.D. Starbuck, "The Intimate Senses as Sources of Wisdom," *Journal of Religion* 1 (1921), 129–45.

36 Thomas Reid, *Essays on the Intellectual Powers of Man*, Essay 2, 16.

37 Ibid.

38 William Drummond of Hawthornden (1623), *The Cypresse Grove*, p. 183.

39 Sigmund Freud (1905), "Three Contributions to the Theory of Sex," *Basic Writings* (New York: Random House, 1938), p. 605.

40 George Byron (1810), quoted by M. Csaky, *How Does It Feel?* (London: Thames and Hudson, 1979).

41 Hardin, *Color for Philosophers*.

42 Ludwig Wittgenstein, *Philosophical Investigations*, 2, 11, trans. G.E.M. Anscombe (Oxford: Blackwell, 1958).

43 Maurice Bowra, *Memories* (Oxford: Oxford University Press, 1967).

44 Andrew Marvell (1681), "The Garden," in *The Metaphysical Poets*, ed. Helen Gardner (Harmondsworth: Penguin, 1957).

45 Wassily Kandinsky, quoted in *How Does It Feel?*, ed. Csaky.

46 See reviews in Patrick Trevor-Roper, *The World Through Blunted Sight* (London: Thames and Hudson, 1970); and in Tom Porter and Byron Mikellides, eds., *Colour for Architecture* (London: Studio Vista, 1976).

47 Porter and Mikellides, *Colour for Architecture*.

48 Kurt Goldstein, "Some Experimental Observations Concerning the Influence of Colors on the Function of the Organism," *Occupational Therapy* 21 (1942), 147–51.

49 L. Halpern, "Additional Contributions to the Sensorimotor Induction Syndrome in Unilateral Disequilibrium With Special Reference to the Effect of Colors," *Journal of Nervous and Mental Diseases* 123 (1956), 334–50.

50 Manfred Clynes, *Sentics: The Touch of Emotions* (London: Souvenir Press, 1977).

51 Samuel Coleridge (1808), *Anima Poetae*, reprinted in *The Poetry of Earth*, ed. E.D.H. Johnson (London: Gollancz, 1966), p. 128.

52 William Wordsworth (1798), "Lines Composed a Few Miles Above Tintern Abbey," *Selected Poems of William Wordsworth*, ed. Roger Sharrock (London: Heinemann, 1958).

53 Plato, *Timaeus*, 47B, in *Philosophies of Beauty*, trans. and ed. E.F. Carritt (Oxford: Clarendon Press, 1931).

54 Giovanni Boccaccio (1358), *Decameron*, quoted by E.H. Gombrich, *Meditations on a Hobby Horse* (London: Phaidon Press, 1963), p. 17.

55 Alain Erlande-Brandenburg, *La Dame à la Licorne* (Paris: Editions de la Réunion des Musées Nationaux, 1978).

56 Wordsworth (1798), "The Tables Turned" and "Expostulation and Reply," in *Selected Poems*, Sharrock.

57 John Constable, quoted by Michael Middleton in *Handbook of Western Painting* (London: Thames and Hudson, 1961).

58 Immanuel Kant (1790), *The Critique of Judgement*, Book 1, 2, in *Philosophies of Beauty*, Carritt.

59 Paul Cézanne, in conversation with J. Gasquet, quoted by Ernest G. Schactel in *Metamorphosis* (London: Routledge and Kegan Paul, 1963).

60 Aldous Huxley, *The Doors of Perception* (New York: Harper and Row, 1954), pp. 25, 19, 20, 41.

61 Quoted in S. Cohen, *Drugs of Hallucination: The Uses and Misuses of LSD* (London: Secker and Warburg, 1964), pp. 167–69.

62 Nicholas Humphrey, "Interest and Pleasure: Two Determinants of a Monkey's Visual Preferences," *Perception* 1 (1972), 395–416.

63 Nicholas Humphrey and Graham Keeble, "Do Monkeys' Subjective Clocks Run Faster in Red Light Than in Blue?," *Perception* 6 (1977), 7–14; "Effects of Red Light and Loud Noise on the Rate at Which Monkeys Sample Their Sensory Environment," *Perception* 7 (1978), 343–48.

64 Nicholas Humphrey and Graham Keeble, "Interactive Effects of Unpleasant Light and Unpleasant Sound," *Nature* 253 (1975), 346–47.

65 Roger Fry (1926), *Transformations*, chap. 1, in *Introductory Readings in Aesthetics*, ed. John Hospers (London: The Free Press, 1969).

66 John Locke (1690), *An Essay Concerning Human Understanding*, Book 2, chap. 1, 5, ed. Peter H. Nidditch (Oxford: Clarendon Press, 1975).

67 Bertrand Russell, *Introduction to Mathematical Philosophy* (London: Allen and Unwin, 1919), p. 71.

68 Locke, *An Essay Concerning Human Understanding*, Book 2, chap. 32, 15.

69 Wittgenstein, *Philosophical Investigations*, 1, 272.

70 Wittgenstein, "Notes for Lectures on 'Private Experience' and 'Sense Data,' " ed. Rush Rhees, *The Philosophical Review* 77 (1968), 284.

71 Denis Diderot (1754), *On the Interpretation of Nature*, 10, 23, in *Diderot: Selected Writings*, trans. J. Stewart and J. Kemp (London: Lawrence and Wishart, 1937).

72 Lewis Carroll (1865), *Alice's Adventures in Wonderland*, chap. 5 (London: Chancellor Press, 1982).

73 I. Kohler, cited by Ronald H. Forgus in *Perception* (New York: McGraw Hill, 1966).

74 Robert B. Welch, *Perceptual Modification* (New York: Academic Press, 1978).

75 Paul Bach-y-Rita, *Brain Mechanisms in Sensory Substitution* (London: Academic Press, 1972).

76 Carroll, *Alice's Adventures in Wonderland*, chap. 6.

77 Macdonald Critchley, *The Parietal Lobes* (London: Hafner, 1966), p. 289.

78 J.M. Oxbury, Susan M. Oxbury, N.K. Humphrey, "Varieties of Colour Anomia," *Brain* 92 (1969), 847–60.

79 Alcock, *Parapsychology: Science or Magic?*, p. 86.

80 A.J. Marcel, "Conscious and Preconscious Perception: Experiments on Visual Masking and Word Recognition," *Cognitive Psychology* 15 (1983), 197–237.

81 M. Eagle, "The Effects of Subliminal Stimuli of Aggressive Content Upon Conscious Cognition," *Journal of Personality* 27, (1959), 578–600.

82 Lawrence Weiskrantz, *Blindsight* (Oxford: Clarendon Press, 1986).

83 Nicholas Humphrey, "Vision in a Monkey Without Striate Cortex: A Case Study," *Perception* 3, (1974), 241.

84 Nicholas Humphrey, "Nature's Psychologists," British Association for the Advancement of Science Lister Lecture, 1977, reprinted in Humphrey, *Consciousness Regained*.

85 Anthony J. Marcel, "Phenomenal Experience and Functionalism," in *Consciousness in Contemporary Science*, ed. Marcel and Bisiach, pp. 121–58.

86 Locke, *An Essay Concerning Human Understanding*, Book 4, chap. 2, 1.

87 William Shakespeare (1595), *Richard II*, 1, 3.

88 Frank G. Burgess, "The Purple Cow," in *Everyman's Dictionary of Quotations and Proverbs* (London: Dent, 1951).

89 Cited by Marcus Raichle, "Images of the Functioning Human Brain," in *Images and Understanding*, ed. H. Barlow, C. Blakemore, M. Weston-Smith (Cambridge: Cambridge University Press, 1990), pp. 284–96.

90 William Shakespeare (1605), *Macbeth*, 2, 1.

91 Samuel Coleridge (1803), Letter quoted by Richard Holmes, *Coleridge: Early Visions* (London: Hodder and Stoughton, 1989), p. 354.

92 Oliver Sacks, *The Man Who Mistook His Wife for a Hat* (London: Duckworth, 1985).

93 Critchley, *The Parietal Lobes*.

94 Nicholas Humphrey, "Contrast Illusions in Perspective," *Nature* 232 (1970), 91–93.

95 Robert H. Thouless, "Phenomenal Regression to the Real Object, II," *British Journal of Psychology* 22, (1931), 1–30.

96 John Donne (1619), "A Hymn to Christ, at the Authors Last Going Into Germany," *Donne: Poetical Works*, ed. Herbert Grierson, (London: Oxford University Press, 1937).

97 Lewis Carroll (1889), *Sylvie and Bruno*, chaps. 5–7 (London: Chancellor Press, 1983).

98 Martha J. Farah, "Is Visual Imagery Really Visual? Overlooked Evidence From Neuropsychology," *Psychological Review* 95 (1988), 307–17.

99 Consciousness workshop convened by Daniel Dennett, Bellagio, May 1990.

100 Aldous Huxley (1936), unpublished speech, quoted in Nicholas Humphrey and Robert Jay Lifton, eds., *In a Dark Time* (London: Faber and Faber, 1984).

101 See for example my discussion in Nicholas Humphrey and G.R. Keeble, "How Monkeys Acquire a New Way of Seeing," *Perception* 5 (1976), 51–56.

102 Samuel Johnson (1776), cited by James Boswell, *Life of Johnson*, vol. 3 (London: Everyman, 1925).

103 Locke, *An Essay Concerning Human Understanding*, Book 3, chap 9, 9.

104 Stuart Sutherland, review of *Consciousness Regained*, *Nature* 307 (1984), 391.

105 Milan Kundera, *Immortality* (London: Faber and Faber, 1991), p. 225.

106 Thomas Traherne (1670), *Centuries of Meditation*, Century 3, 3 (London: Dent, 1908).

107 Jean-Jacques Rousseau (1754), *A Discourse on Inequality*, trans. Maurice Cranston (Harmondsworth: Penguin, 1984), p. 109.

108 Ray Jackendoff, "Is There a Faculty of Social Cognition?," unpublished manuscript, 1989.

109 Nicholas Humphrey (1975), "The Social Function of Intellect," reprinted in Humphrey, *Consciousness Regained*.

110 Daniel Stern, *The Interpersonal World of the Infant* (New York: Basic Books, 1985), p. 78.

111 Eduardo Bisiach and Giuliano Geminiani, "Anosognosia Related to Hemiplegia and Hemianopia," in *Awareness of Deficit After Brain Injury*, G.P. Prigatano and D.L. Schacter, eds. (New York: Oxford University Press, 1990).

112 Eduardo Bisiach, "Language Without Thought," in *Thought Without Language*, L. Weiskrantz, ed. (Oxford: Clarendon Press, 1988), pp. 464–91.

113 William Shakespeare (1605), *Othello*, 3, 324.

114 Gerard Manley Hopkins, *The Starlight Night* (1918).

115 Wilfred Sellars, *Science, Perception and Reality* (London: Routledge and Kegan Paul, 1963).

116 Edward Titchener (1896), quoted by E.G. Boring, *Sensation and Perception in the History of Experimental Psychology* (New York: Appleton-Century-Crofts, 1942), p. 10.

117 D.J. McFarland, *The Encyclopedic Dictionary of Psychology*, ed. Rom Harré and Roger Lamb (Oxford: Blackwell, 1983), p. 448.

118 William Blake (1810), A Vision of the Last Judgement, Descriptive Catalogue, in The Complete Writings of William Blake, ed. Geoffrey Keynes (Oxford: Oxford University Press, 1957).

119 William Blake (1818), The Everlasting Gospel, d, 1, 103., in ibid.

120 Ronald Melzack, The Puzzle of Pain (Harmondsworth: Penguin, 1973), p. 50.

121 Ambroise Paré (1552), quoted in ibid., p. 50.

122 Cited by J.M. Heaton, The Eye: Phenomenology and Psychology of Function and Disorder (London: Tavistock Publications, 1968), p. 184.

123 T.S. Eliot (1936), "Burnt Norton," Four Quartets (London: Faber and Faber, 1946).

124 Daniel Dennett and Marcel Kinsbourne, "Time and the Observer: The Where and When of Consciousness in the Brain," Brain and Behavioral Sciences (forthcoming).

125 Ronald Melzack and Howard Eisenberg, "Skin Sensory Afterglows," Science 159 (1968), 445–47.

126 Kundera, Immortality, p. 225.

127 Philip Steadman, The Evolution of Designs (Cambridge: Cambridge University Press, 1979), chap. 7.

128 Alan Turing's original paper, "Computing Machinery and Intelligence" (1950), together with some of the discussion it has spawned, such as John Searle's "Mind, Brains, and Programs" (1980), are reprinted in The Mind's I, ed. Douglas R. Hofstadter and Daniel C. Dennett (London: Harvester Press, 1981).

129 Frank Jackson, "What Mary Didn't Know," Journal of Philosophy 83 (1986).

130 Colin McGinn, "Could a Machine be Conscious?," in Mindwaves, ed. Blakemore and Greenfield, p. 287.

131 Thomas Nagel, The View From Nowhere (New York: Oxford University Press, 1986), p. 47.

132 Robert van Gulick, "A Functionalist Plea for Self-Consciousness," The Philosophical Review 97 (1988), 149–81.

133 Thomas H. Huxley, Lessons in Elementary Physiology, 8 (1866), 210.

134 Saul Kripke, "Identity and Necessity," in Identity and Individuation, ed. M. Munitz (New York: New York University Press, 1971).

135 Encyclopaedia Britannica, 14th ed. 1929.

136 Thomas Gray (1750), "Elegy Written in a Country Churchyard," in The New Oxford Book of English Verse, ed. Helen Gardner (Oxford: Oxford University Press, 1972).

INDEX

Nicholas Humphrey is a distinguished theoretical psychologist who has held research and teaching posts at both Oxford and Cambridge, as well as fellowships in the United States and Germany. He is well known as a writer and television documentary maker, whose work includes the famous Bronowski Memorial Lecture on nuclear weapons: "Four Minutes to Midnight," and books such as *Consciousness Regained, In a Dark Time,* and *The Inner Eye.* His interests are wide ranging: he studied mountain gorillas with Dian Fossey in Rwanda, has made important discoveries about the brain mechanisms underlying vision, proposed the now-celebrated theory of the "social function of human intellect," and is the only scientist ever to edit the literary journal *Granta.* He has been the recipient of several honors, including the Martin Luther King Memorial Prize in 1985. He divides his life between London and Cambridge, where he is currently a senior Research Fellow of Darwin College.